LIVING IMAGES

LIVING IMAGES

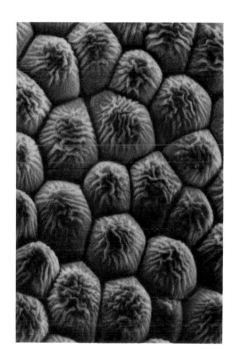

Biological Microstructures Revealed by
Scanning Electron Microscopy

Gene Shih
Richard Kessel

SCIENCE BOOKS INTERNATIONAL
BOSTON, MASSACHUSETTS

Editorial offices: Science Books International, Publishers, 30 Granada Court, Portola Valley, CA 94025.
Sales and customer service offices: 51 Sleeper Street, Boston, MA 02210.

Library of Congress Cataloging in Publication Data

Shih, Ching Yuan, 1934–
    Living images.

    Bibliography: p.
    1. Ultrastructure (Biology) 2. Scanning
electron microscope. I. Kessel, Richard G.,
1931– . II. Title.
QH212.S3S54 1982   574'.022'2   82-10431
    ISBN 0-86720-006-5   ISBN 0-86720-008-1 (pbk)

Publisher Arthur C. Bartlett
Book and cover design Paula Schlosser
Production Bookman Productions
Composition Typothetae
Printer and binder Halliday Lithograph

Printed in the United States of America

Printing number (last digit)   10   9   8   7   6   5   4   3   2   1

CONTENTS

PREFACE

FOR MORE THAN twelve years we have utilized the scanning electron microscope to study biological structures, and during that time we have accumulated a number of photographs that are breathtaking in their display of biological detail. This book is a collection of selected images that we would like to share with our readers.

Although one picture may be worth a thousand words, we have also included a simple explanation for each picture or group of pictures in order to provide a more complete understanding of the presentation. The illustrations are grouped so that the reader can proceed from the relatively "simple" microorganisms—with their marvelous symmetry—to more complex organisms such as plants and animals and parts of these organisms.

We would like to thank our book designer, Paula Schlosser, whose careful arrangement of the pictures has brought out the beauty of the individual micrographs. The text was simplified from our original version by the skill of Dr. David Freifelder, so that readers with a minimum of scientific background can appreciate the significance of complex biological microstructures.

The authors would also like to thank Professor Emeritus H. W. Beams for his constant encouragement during our twelve years of work with the scanning electron microscope, and our colleagues in various departments at the University of Iowa for providing us with some of the materials. Many of the animal tissue and organ micrographs were obtained from materials prepared by Dr. Randy Kardon, and his efforts are very much appreciated. Finally, we would like to thank our publisher, Arthur Bartlett, and the project coordinator, Hal Lockwood, for their effort in bringing this book to the public.

G. C. Shih
R. G. Kessel

INTRODUCTION

IT HAS BEEN SAID that if you look at anything closely enough, it will appear interesting. This is certainly true of the living world. A drop of water can be seen to be teeming with life and the microscopic cells contained therein often possess surprising, complex, and often aesthetically pleasing forms. The tiny insect is seen to have a large number of efficiently designed parts. A lovely flower petal, on close observation, is covered with millions of tiny geometrical figures. The lowly worm is wrinkled and its wrinkles have wrinkles; and these are wrinkled even further. In this book we show you some of the marvelous forms that have been seen with a scanning electron microscope. This display has two functions—to provide what can be an aesthetic experience (simply by looking at the pictures) and to learn a little about how the biological world functions (by reading the captions).

Various microscopic devices are available to observe and display the microstructures of the world around us. Each of these devices has advantages and limitations. For example, of the three major types of microscopes—the light microscope, the transmission electron microscope, and the scanning electron microscope—only the light microscope allows one to examine living material; the two types of electron microscopes require that any sample to be observed be dry and viewed in a vacuum. On the other hand, the limit of resolution—that is, the minimum separation between two objects that allows them to be seen as two objects and not one object—of the light microscope is about 0.00025 millimeter, or 2500 angstrom units. In practice this means that the light microscope has a maximum effective magnification of about 1200.

The transmission electron microscope (the most common kind of electron microscope) has been widely used in biology and medicine for the past thirty years. With this microscope an electron beam, rather than visible light, is used as a source of illumination and, for strictly physical reasons, this provides a limit of resolution of 0.0000005 millimeter, or 5 angstrom units—500-fold greater than with the light microscope. This means that samples can be magnified about one million times. With such a microscope individual molecules can be seen and the molecular detail seen in biological specimens is extraordinary. A disadvantage of the transmission electron microscope is that it can only accept a very small specimen and furthermore the sample must be exceedingly thin; if it is not naturally thin, it must be sliced in order that the electrons can pass through the sample. Furthermore, it is not possible to use a magnification of less than about 1000.

The scanning electron microscope used to obtain the pictures shown in this book bridges the magnification gap, for it can be used in a magnification range of 10 to 100,000. It also yields an almost three-dimensional view not obtained with the standard electron microscope. On the other hand, the limit of resolution of the scanning electron microscope is only about 50 to 100 angstrom units; this is, however, adequate for all except the most detailed examination of biological structures. A unique feature of the scanning electron microscope is its tremendous depth of focus—this means that most of the sample can be in sharp focus simultaneously. This is especially valuable when studying samples at low magnification; in contrast, with a light microscope, the depth of focus is about 500-fold less and except for perfectly flat samples, a significant part of the sample is always out of focus. Another advantage of the scanning electron microscope is that very large samples can be placed in the specimen chamber—considerably larger than that which can be accepted by either the light microscope or the transmission electron microscope. Furthermore, the specimen can be tilted so that it can be viewed at different orientations. This, plus the depth of focus just mentioned, gives one a beautiful three-dimensional perspective of the microscopic world. However, it is clear that the maximum amount of information about the biological specimen is obtained by observing the sample with all three types of microscope.

LIVING IMAGES

BACTERIA

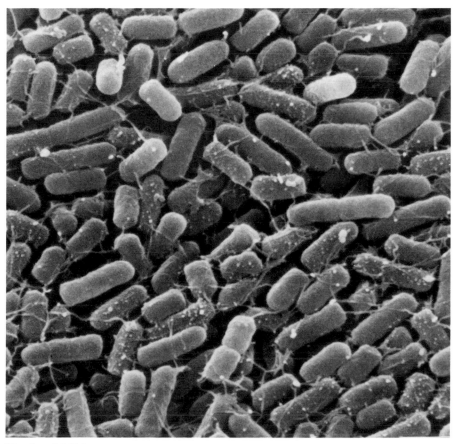

Fig. 1 × 1/4

Bacteria are living cells consisting of a rigid wall surrounding a membranous sac containing about 10,000 different chemicals. A mass of about 100 billion cells are growing on the agar surface in the test tubes shown in Fig. 1. Bacteria grow rapidly and when reaching a certain size, split in two, giving rise to identical twins. When there is an adequate food supply, the life cycle takes about 20 minutes. Although many bacteria cause disease, most perform useful functions. For example, bacteria decompose natural and artificial waste, produce soil, and are used in the manufacture of all dairy products. Also, many important industrial chemicals are synthesized by controlled bacterial growth and in recent years, genetically engineered bacteria have been used to produce drugs and biologically active substances. Bacteria come in many shapes: cocci are spherical (Fig. 3), bacilli are rod-like (Figs. 2, 4, 6), and spirilla are spirals (Fig. 5).

Fig. 2 × 14,700

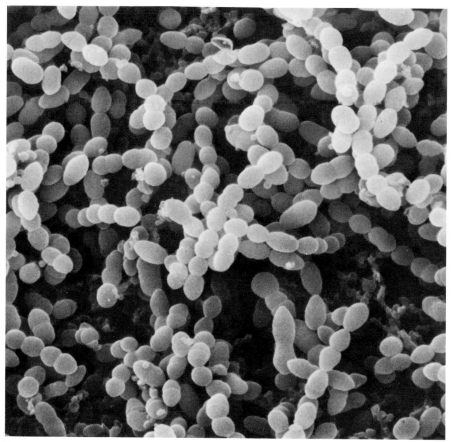

Fig. 3 × 8,820

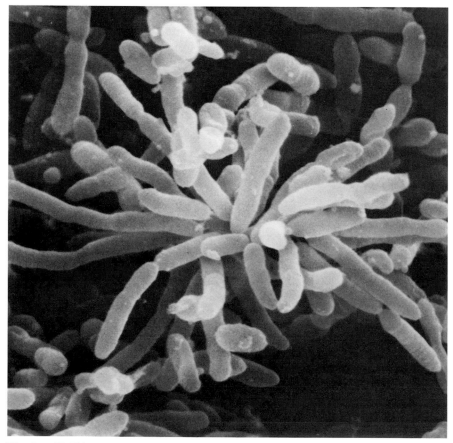

Fig. 4 × 11,350

2

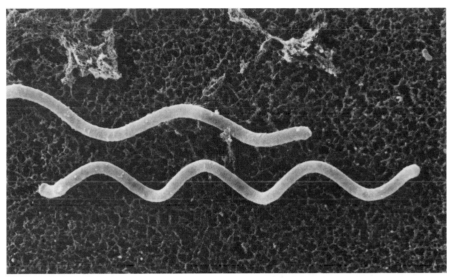

Fig. 5 × 3,300

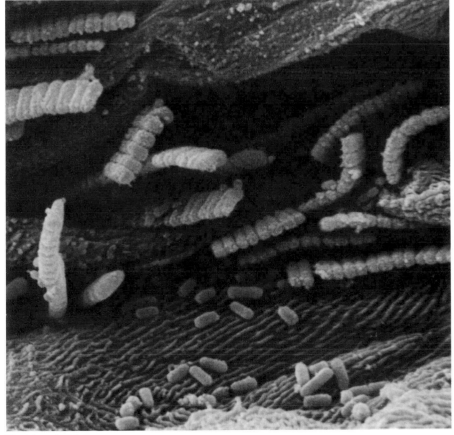

Fig. 6 × 6,700

BACTERIA
Dental plaque

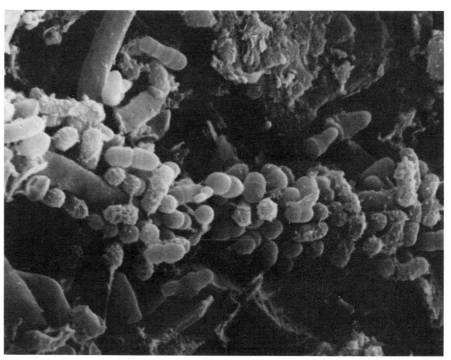

Fig. 1 × 1.5

Fig. 2 × 12,400

Bacteria are present on human teeth (Fig. 1) in dental plaque, a pasty material found on the tooth surface, especially at the gum line. Plaque consists of a variety of bacteria mixed with food particles and a sugar-like adhesive which the bacteria make. The bacteria feed on the particles, making acid, which dissolves the tooth. The bacteria may be spherical (Fig. 2), rod-like (Fig. 3), or filamentous (Fig. 4). The particular kinds of bacteria present vary from person to person and from tooth to tooth. Among the bacteria present in plaque, probably the most significant is *Streptococcus mutans* (Fig. 2), which has been found in all cavities and is thought to be the bacterium that starts cavity formation by localized secretion of acid. The keys to the reduction of cavities are fluoride intake to make the tooth more resistant to bacterial acid, and vigorous brushing after eating, which disperses the bacteria and prevents the accumulation of high local concentrations of acid.

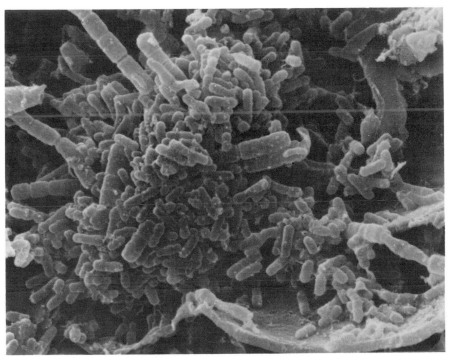

Fig. 3 × 7,500

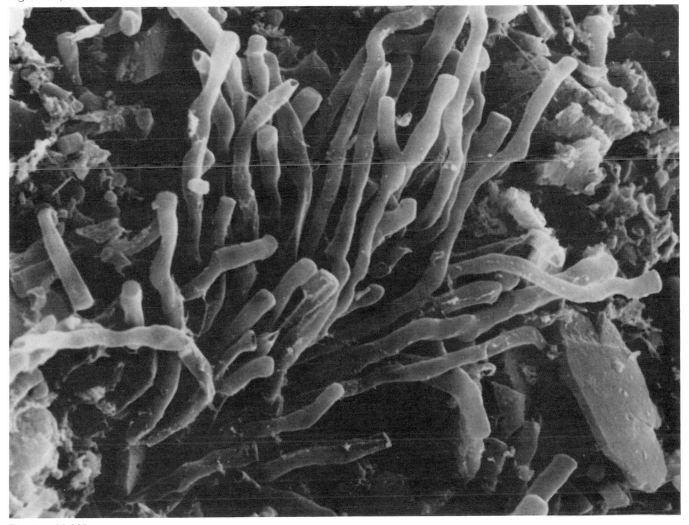

Fig. 4 × 10,800

BACTERIA
Nitrogen-fixing bacteria

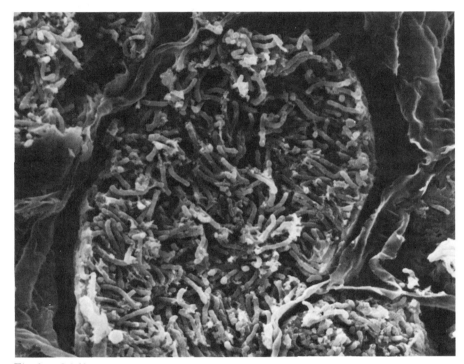

Fig. 1 × 1/2

Fig. 2 × 2,360

The roots of leguminous plants such as peas, beans, and clover are covered with visible clusters of cells which together form what is called a root nodule (Fig. 1). The bacterium *Rhizobium* is one of several bacteria that live in these nodules. The bacteria obtain nutrients from the plant and in return make nitrogen from the atmosphere available to the plant, by a process called *nitrogen fixation*. When two organisms, in this case, *Rhizobium* and the plant, mutually support one another, they are said to be *symbiotic*. Particular species of bacteria are found in different plants. This is illustrated by the presence of bacteria of different shapes in the nodules. Branched bacteria are seen in the nodules of the pea (Fig. 3), egg-shaped bacteria in the white clover (Fig. 4), and rod-like bacteria in beans (Fig. 5). In order to form a nodule, the bacteria must first adhere to specific sites on the hairs of the root tip. Not all plants have these sites. Genetic engineering techniques are being applied to both plants and bacteria to make compatible systems. For example, if plants other than legumes (for example, wheat and corn) could acquire root nodules, nitrogenous fertilizers would no longer have to be applied to grow these plants.

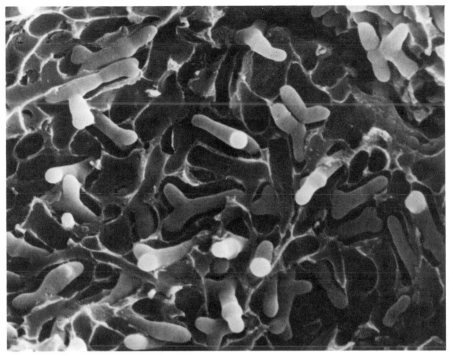

Fig. 3 × 7,390

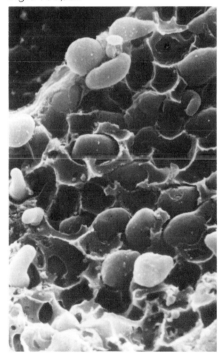

Fig. 4 × 4,880

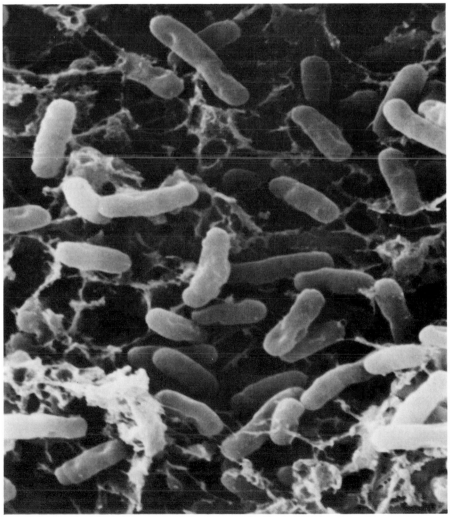

Fig. 5 × 10,580

ALGAE

Euglena

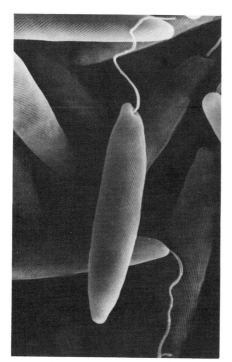

Fig. 1 × 1,790

Algae are single, free-living cells differing from bacteria in that they contain chlorophyll and are capable of photosynthesis. *Euglena* (Fig. 1) is an alga widely distributed in fresh-water ponds. Its long oval body has a small hole at one end used for taking in microscopic food particles. Adjacent to this "mouth" is a hair-like *flagellum* which whips back and forth and thereby propels the cell through the water. A higher magnification photograph (Fig. 2) shows that the flagellum emerges from a depression and that the cell wall of *Euglena* consists of strips arranged in a gentle spiral pattern on the cell surface.

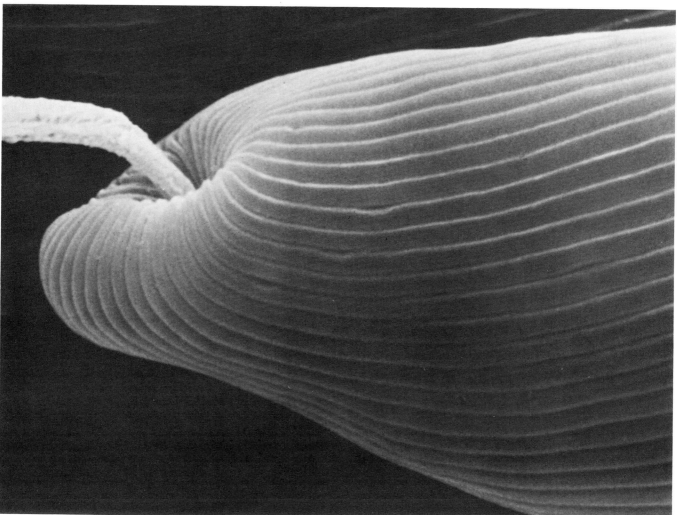

Fig. 2 × 25,000

ALGAE

Chlamydomonas

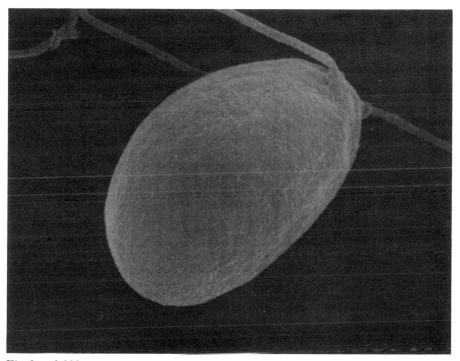

Fig. 1 × 1,545

Fig. 2 × 9,000

Chlamydomonas (Fig. 1) is one of the most common forms of green algae in nature, found in most fresh-water pools and in soil. Two whip-like flagella at the rear of the cell (shown in the higher magnification photo of Fig. 2) enable it to move through water. *Chlamydomonas* needs little to grow other than water, carbon dioxide, a few simple and common salts, and light, for it is able to make its food by photosynthesis. It is thought that *Chlamydomonas* is structurally similar to the ancestral organism from which modern plants had their origin. *Chlamydomonas* is able to reproduce both asexually (by simple cell division) and sexually (by the merging of two cells followed by two cycles of cell division). *Chlamydomonas* is a valuable research tool for understanding the genetic properties of sexually reproducing microorganisms.

ALGAE
Desmids

Fig. 1 × 730

Desmids are widely distributed green algae having an aesthetically pleasing symmetry. Each cell consists of two half-units (Fig. 1), which are mirror images of one another. In its asexual mode of reproduction the halves separate and each half regenerates a mirror image. In the sexual mode of reproduction two desmids come in contact and merge to form a single large unit. This unit later breaks open, releasing four spherical particles, which, in time, grow and change their shapes and become the symmetrical desmid cell. Some common shapes of desmids are shown in Figs. 2 through 4.

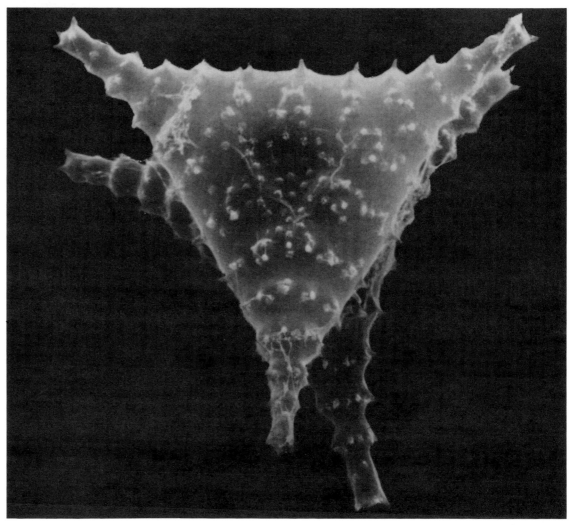

Fig. 2 × 3,400

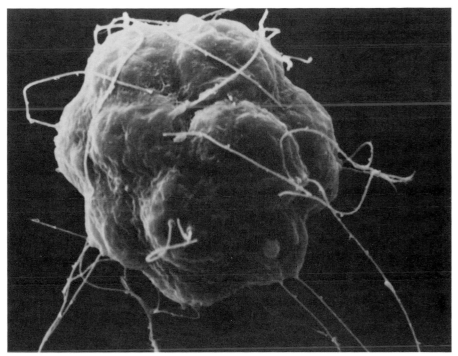

Fig. 3 × 3,800

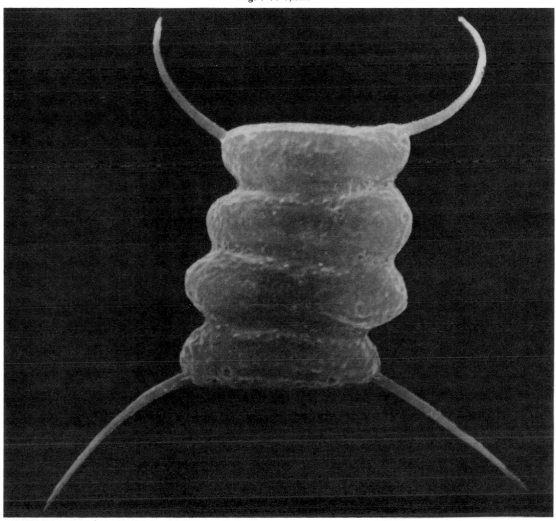

Fig. 4 × 3,435

ALGAE

Diatoms

Fig. 1 × 70

The structural complexity and intricate design of diatoms make them the beauties of the algal world (Figs. 1–13). The design is built into a stony cell wall that consists of two overlapping portions fitted together like two halves of a box. Diatoms are found in both fresh and salt water. When they die, the stony wall sinks to the bottom of the ocean or lake where it accumulates as sediment. The walls persist for millions of years and have in time become compacted into layers called *diatomaceous earth*. This mineral material is used commercially in making insulating bricks, as a filter, and as an abrasive in toothpaste.

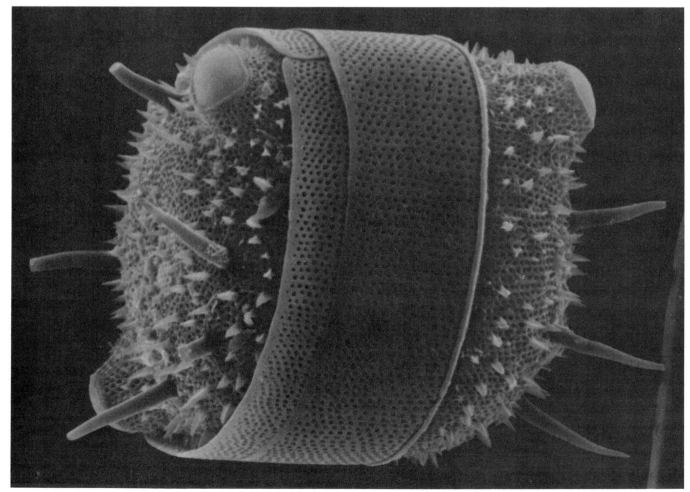

Fig. 2 × 1,950

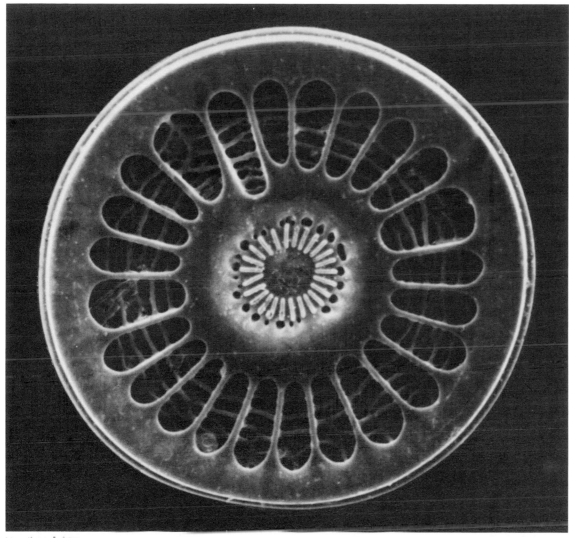

Fig. 3 × 1,080

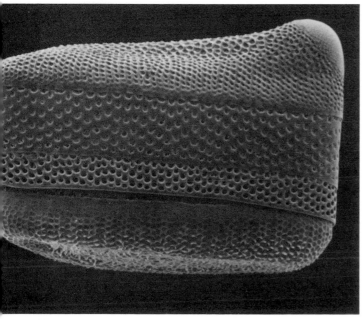

Fig. 4 × 325

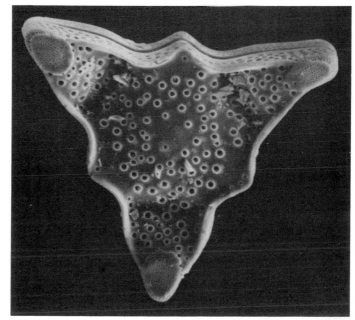

Fig. 5 × 860

13

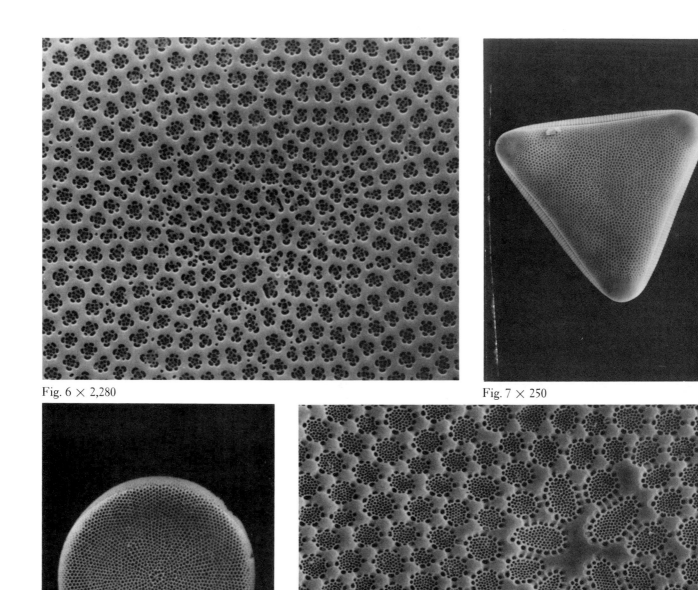

Fig. 6 × 2,280

Fig. 7 × 250

Fig. 8 × 330

Fig. 9 × 2,340

14

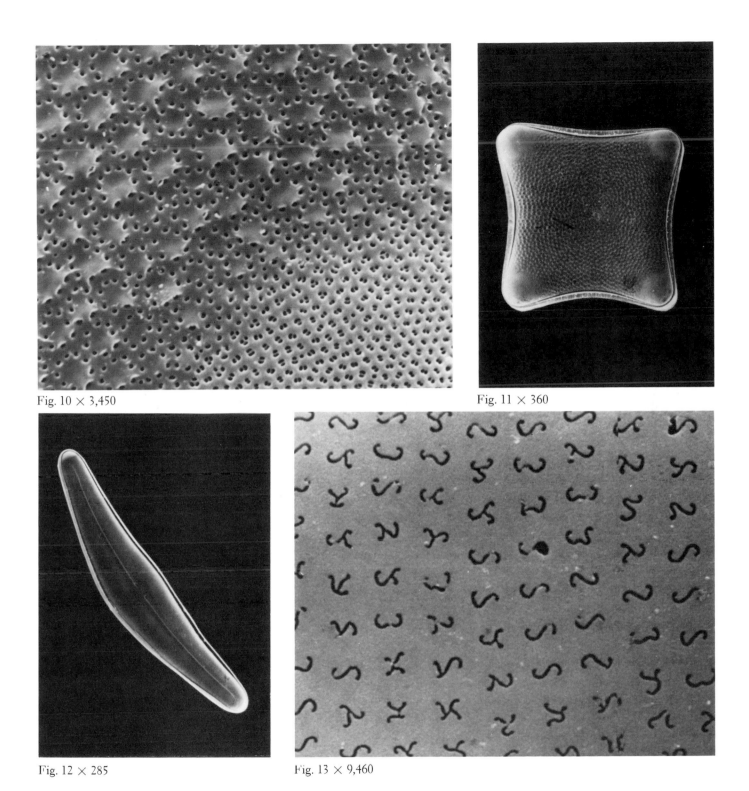

Fig. 10 × 3,450

Fig. 11 × 360

Fig. 12 × 285

Fig. 13 × 9,460

SLIME MOLD

Physarum

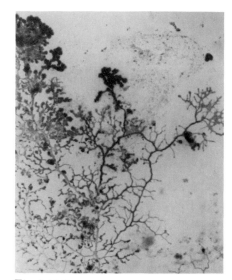

Fig. 1 × 2

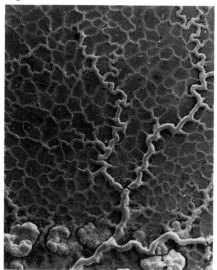

Fig. 2 × 11

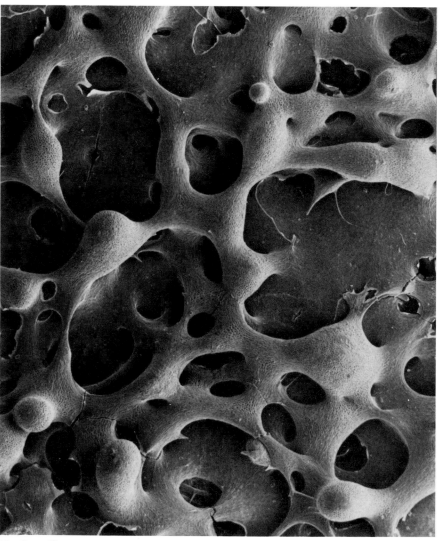

Fig. 3 × 60

Physarum is a slime mold found on damp soil, rotting logs, and decaying leaves, where it looks like glistening yellow slime. A close look shows that *Physarum* is a mass of branches called *plasmodia* (Fig. 1). The plasmodia move toward tiny particles of decayed organic matter used as food. The tip of the plasmodium flows around the particle and engulfs it. If the environment of the slime mold becomes depleted of moisture, the entire mold becomes dried out (Fig. 2) and remains dormant until water is restored. A close-up view of the dried plasmodial surface is shown in Fig. 3.

YEAST

Saccharomyces cerevisae

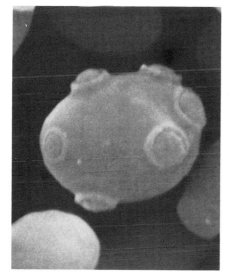

Fig. 1 × 1/3

Yeasts have served humanity for centuries in making wine and bread (Fig. 1). Dried yeast is put in vitamin pills as a source of B vitamins. Yeasts are also used industrially as they can synthesize a large number of valuable substances. Yeasts differ from other molds in that they are free living single cells and can grow in both liquid or on surfaces using carbohydrates for food. Yeast reproduce by formation of a small surface bubble called a *bud* (Fig. 3). The bud enlarges until it reaches the size of its parent and then separates from the parental cell, leaving a small circular scar—a bud scar (Fig. 2)—on the parental cell.

Fig. 2 × 12,000

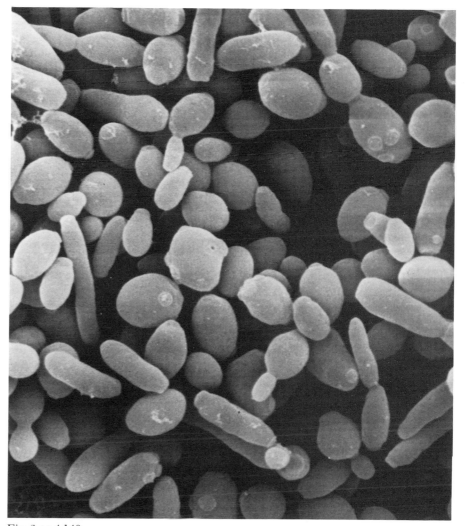

Fig. 3 × 4,140

17

BREAD MOLD

Rhizopus stolonifer

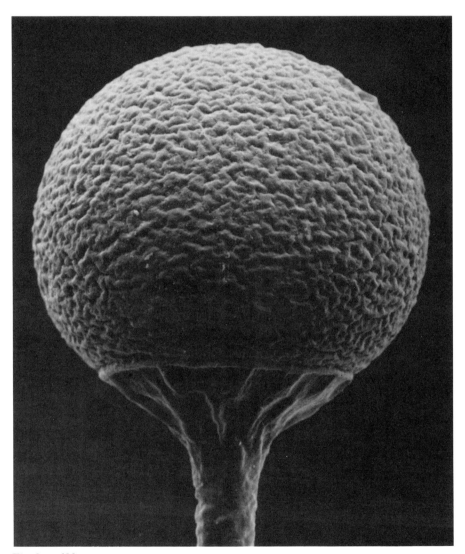

Fig. 1 × 42

Fig. 2 × 680

Rhizopus is the common black mold of bread (Fig. 1) but it also grows in soil, on fruit, and on all kinds of decaying material. This fungus grows by sending out rootlike threads that penetrate the substance on which it grows. The threads seek out and absorb nutrients (Fig. 1). As the fungus matures, a beautiful spherical body (a *sporangium*) forms at the tip of many threads (Fig. 2). These sporangia, which are black and give the mold its color, contain the spores needed to grow new molds. Fig. 3 shows a mature sporangium which is shedding its spores. A close-up view of the spores is shown in Fig. 4.

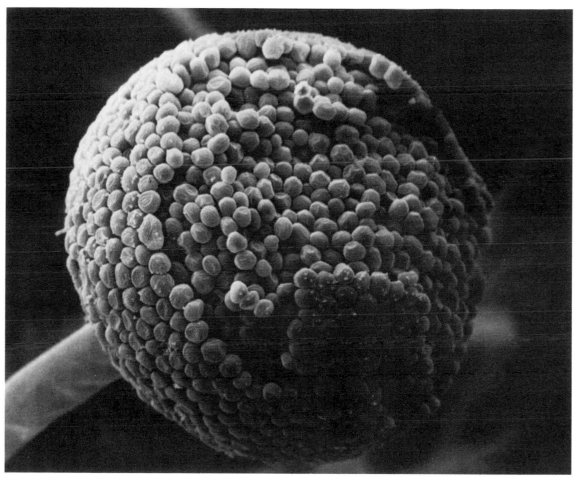

Fig. 3 × 700

Fig. 4 × 5,740

MOLD

Thamnidium elegans

Fig. 1 × 90

Thamnidium grows in both damp soil and manure. It reproduces by forming spores (the "seed" of fungi) on two bodies, both shown in Fig. 1. The long stalk is terminated by a *sporangium*, the tip of which is shown at higher magnification in Fig. 2. When this matures, it becomes sticky and is attached to passing animals which carry it to a new location where the spores can germinate. The short branches are terminated by beautiful clusters of bodies called *sporangiola* (Fig. 3). At maturity the sporangiola burst open, releasing spores which are carried by the wind.

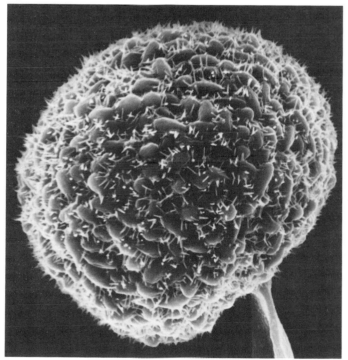

Fig. 2 × 2,900

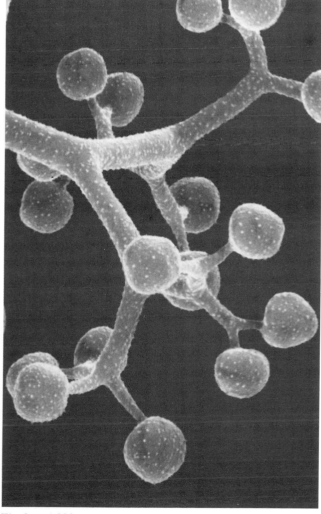

Fig. 3 × 4,080

21

Fig. 1 × 1/2

FUNGI

Penicillium

Penicillium is a mold found on all kinds of decaying material, especially citrus fruits (Fig. 1). Of greatest importance is its ability to synthesize the antibiotic penicillin, which kills bacteria by preventing them from synthesizing a cell wall. The frequent exposure of humans to *Penicillium* is the major cause of penicillin allergy. The filaments of *P. notatum* terminate with segmented branches at the tip of which form spherical spores (Fig. 2). The spores detach and each one is capable of forming a new organism. In some species the spores have small spines and are in elegant array (Fig. 3).

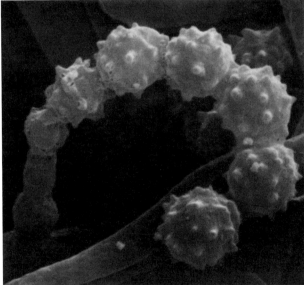

Fig. 3 × 7,200

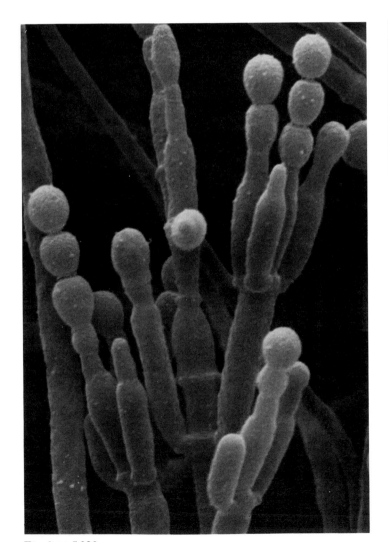

Fig. 2 × 5,320

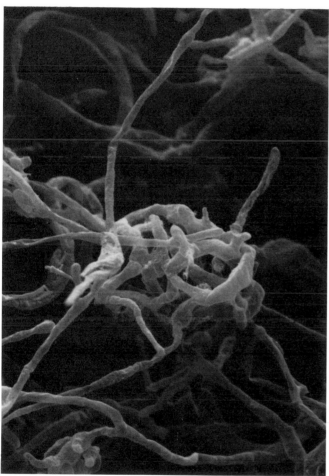

Fig. 1 × 480

FUNGI

Neurospora

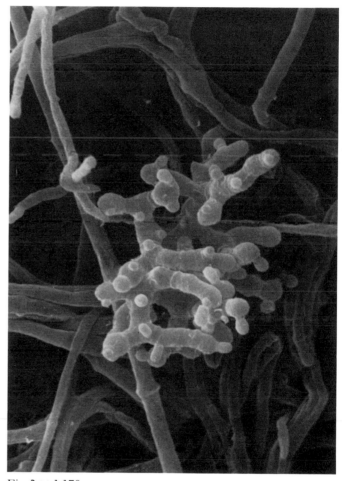

Fig. 2 × 2,730

Fig. 3 × 1,170

Neurospora is a red-orange fungus occasionally found on bread and fruit. Its greatest importance has been in the study of genetics. It grows very rapidly, forming a loose network of strands that may be several inches long. These strands, shown in Fig. 1, are called *mycelia*. *Neurospora* reproduces both sexually and asexually. In the sexual mode collections of spore sacs (*asci*) form in clusters (*conidia*) in a specialized sheath shown in Fig. 2. At maturity the asci burst, releasing the spores. These spores generate a new system, which is either male or female, that reproduces asexually. In the asexual mode, spores are formed on the aerial threads shown in Fig. 3. If a male and female come into contact, portions of the aerial threads fuse and from this fusion there emerges a new organism on which the spores will be produced.

FUNGI

Sordaria

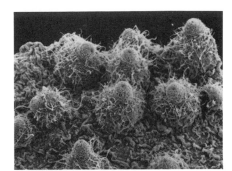

Fig. 1 × 60

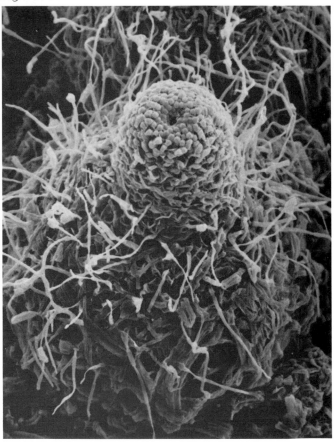

Fig. 2 × 415

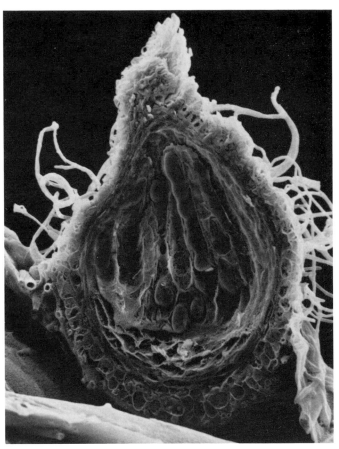

Fig. 3 × 490

Sordaria is a whitish mold that grows as a collection of upright conical masses, shown in Fig. 1. The spore sacs (*asci*) are contained in a globular cluster which is surrounded by a gelatinous sheath (Fig. 2). Each ascus matures separately and, in so doing, it enlarges and eventually protrudes through an opening in the sheath (Fig. 3). Once the spores are released, the ascus collapses and a second ascus matures and enlarges through the sheath. During the maturation process, the mold appears dusty black because of the dark color of the asci.

MUSHROOM

Agaricus bisporus

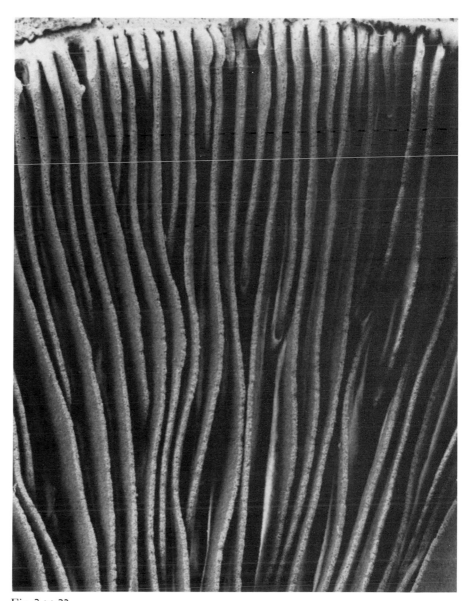

Fig. 1 × 1/2

Agaricus is the white mushroom commonly sold in supermarkets (Fig. 1). Growth begins underground as a mass of fibers (*mycelia*). As the mushroom develops, specialized spore-producing branches aggregate to form small buttons. As the button enlarges, it emerges from the ground. Two types of mushroom, gill formers and spore formers, are distinguished by the structure under the cap. This mushroom has gills (Fig. 2) arranged radially from the stem.

Fig. 2 × 23

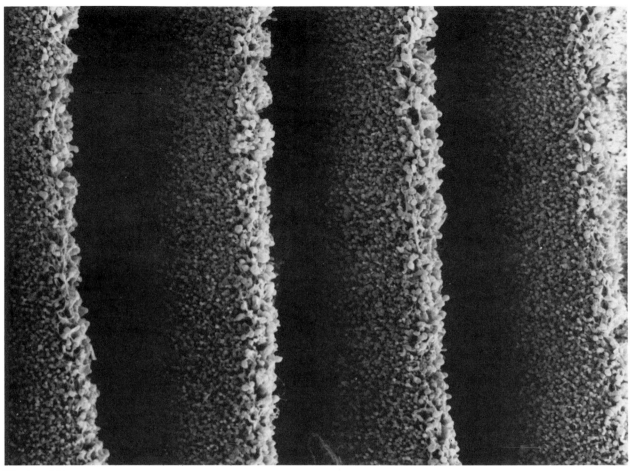

Fig. 3 × 150

Several gills are enlarged in Fig. 3. Note the enlarged cells on the surface of the gill. These cells, called *basidia*, are responsible for forming spores. Individual spores are formed several days after the mushroom has emerged from the ground. These are enlarged in Fig. 4. When mature, the spores fall from the gills onto the ground, where they are dispersed by insects and water. During the few days of spore formation a single mushroom can release more than 10 billion spores.

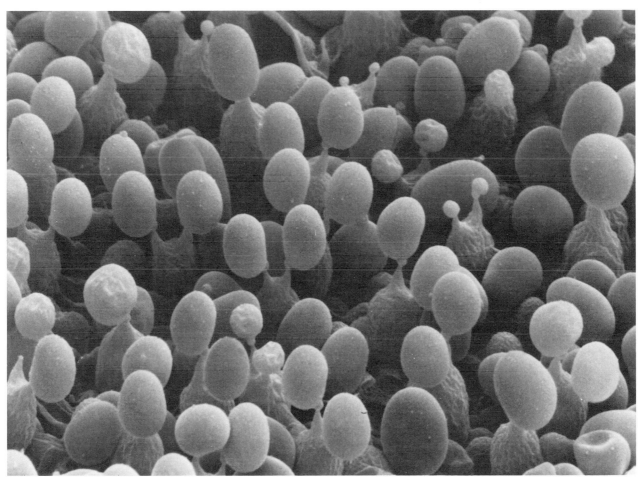

Fig. 4 × 3,150

LIVERWORTS

Marchantia

Fig. 1 × 2

Liverworts are small (1/2 inch), flat green plants found in moist areas (Fig. 1). They are Nature's "pioneers"; on wet rocky surfaces they secrete weak acid, dissolving the rock and converting it to useful soluble minerals, which provide nutrients for lichens and mosses. During dry periods many of the plants on the rocky surface disintegrate, initiating soil formation. The liverworts reproduce in a two-stage cycle, asexually and sexually. In the asexual step they form a "leaf" (Fig. 1) on which propagating structures form (Fig. 2). When the structure is mature (Fig. 3), reproductive cells form in these cups (lower right) and, when released, develop into new plants. In the sexual step male and female "umbrellas" (Fig. 1) develop in which either a sperm or an egg form. Environmental moisture (e.g., dew) provides a medium in which the sperm can swim to the egg. The fertilized egg then develops into a spore-forming organ which releases spores to the environment. The spores form new plants, starting the cycle anew.

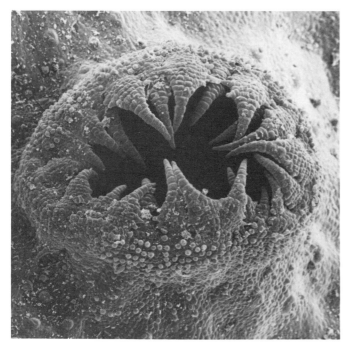

Fig. 2 × 123

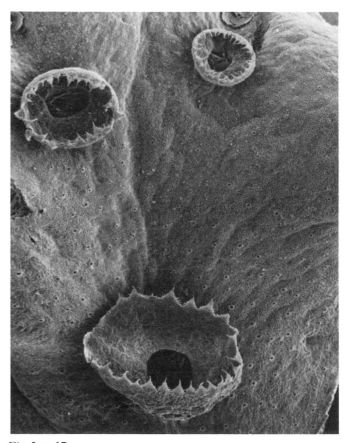

Fig. 3 × 17

MOSSES

Fig. 1 × 1

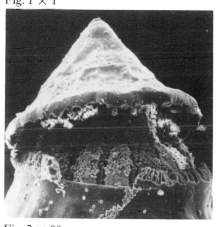

Fig. 2 × 90

Mosses are familiar features of woodlands and shaded stream banks, often forming dark green cushions between moist stones and in damp crevices. Mosses reproduce in the following way. Initially a spore develops to form the familiar green plant (Fig. 1), which consists of a stem, a rootlike system and leaves. At maturity male and female organs are produced at the tip of the stem. Sperm and eggs develop in these organs and fertilization occurs at the tip of the stem. The fertilized egg develops into an upright and quite visible red-brown, spore-producing unit consisting of a foot, which is anchored to the green stem of the parent moss, a brown stalk, and a capsule. These units cover the green moss blanket, giving the entire system a brownish appearance. Spores form in the capsule, whose lid (Fig. 2) falls off at maturity. Under the lid is a set of teeth attached to the capsule wall (Fig. 3). In dry weather the teeth open and release the spores (Fig. 4), which germinate another green plant if moisture is available.

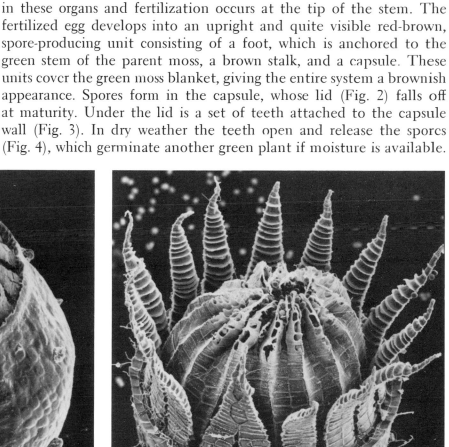

Fig. 3 × 95

Fig. 4 × 124

FERNS

Boston Fern

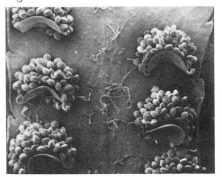

Fig. 1 × 1

Fig. 2 × 12

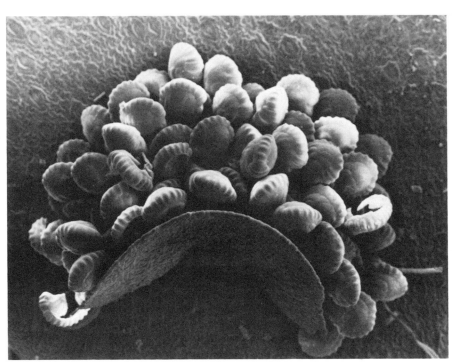

Fig. 3 × 72

Ferns are graceful plants mostly found in moist shady places. They range in size from small aquatic forms a few inches high to gigantic tree ferns. Except for the tree ferns for which the stem forms the trunk, the stem remains underground so that only the leaves are visible. The leaves (*fronds*) are usually extensively divided (Fig. 1) and form graceful clusters on the forest floor. Ferns are spore-formers and the spores are located on the underside of the frond in spore cases which form clusters known as *sori* (Fig. 2). A magnified view of a sorus is shown in Fig. 3. Each spore case releases spores (Fig. 4) which are dispersed by the wind and germinate to form tiny, heart-shaped plants (Figs. 5, 6) on which distinct male and female organs are located. Sperm are released and the sperm swims to the egg on a thin layer of moisture usually on the plant. Fertilization of the egg occurs on this small plant and the fertilized egg forms a tiny frond (Fig. 7). More fronds eventually grow and a mature plant develops.

Fig. 4 × 325

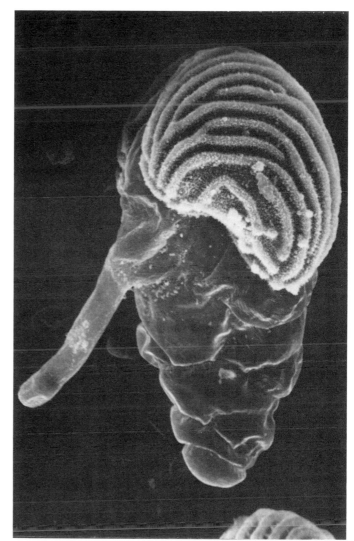

Fig. 5 × 620

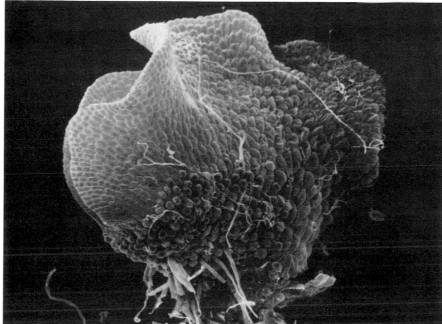

Fig. 6 × 65

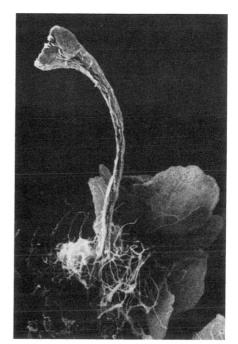

Fig. 7 × 23

31

WATER FERN

Marsilea

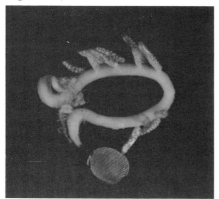

Fig. 1 × 1/4

Fig. 2 × 1

Marsilea is a water fern and is shaped something like a four-leafed clover (Fig. 1). Like the ferns just described, *Marsilea* has both a spore-forming stage and a sexual stage in its life cycle. Each germinating spore (Fig. 2) forms a plant on which male and female organs develop. The sperm are bizarre, consisting of a large sac and a tightly wound spiral band which contains the genetic material (Figs. 3, 4) and is surrounded by about 100 motile hairs called *flagella*. Fig. 3 is an enlargement of one sperm, looking into the spiral band. The flagella propel the sperm toward the egg, which once fertilized, develops into a small frond to begin the spore-forming cycle.

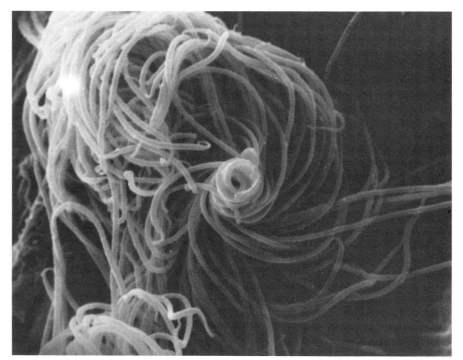

Fig. 3 × 3,000

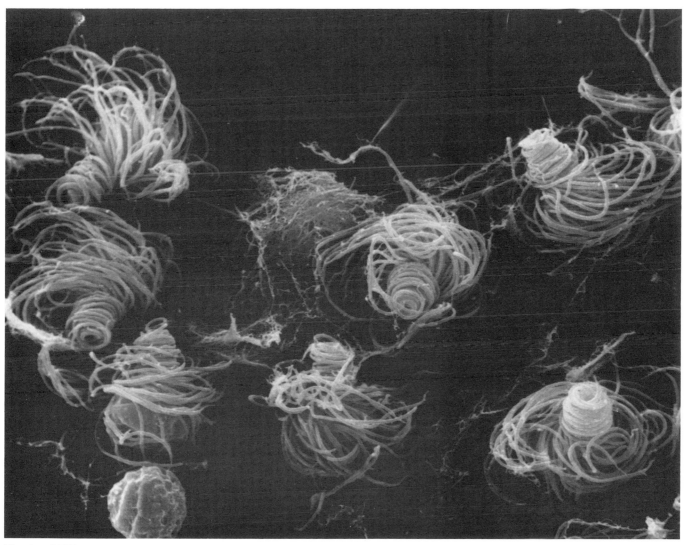

Fig. 4 × 2,720

Fig. 1 × 1

AZOLLA

Green Manure

Azolla is a tiny, fast-growing water fern (Fig. 1) that is used as "green manure" to fertilize rice paddies in Asia. (The term green manure refers to a natural plant that enriches the fertility of soil and thereby avoids the necessity of chemical fertilizers.) The plant consists of branching stems that are densely covered with multi-lobed leaves enlarged in Fig. 2. (Note the lovely sculpturing on the leaves.) One lobe is submerged and the other remains above the surface of the water. The exposed lobe contains numerous cavities (Fig. 3) which house colonies of a particular species of blue-green algae. Chains of these algae are enlarged in Fig. 4. These algae can convert atmospheric nitrogen to compounds which are nutritious for other plants. Thus, by growing *Azolla* in the paddies and then plowing the ferns into the soil, the fertility of the soil is increased. It is believed that this fern may some day play a key role in increasing the world food supply, as it grows easily and rapidly in any fresh water.

Fig. 2 × 46

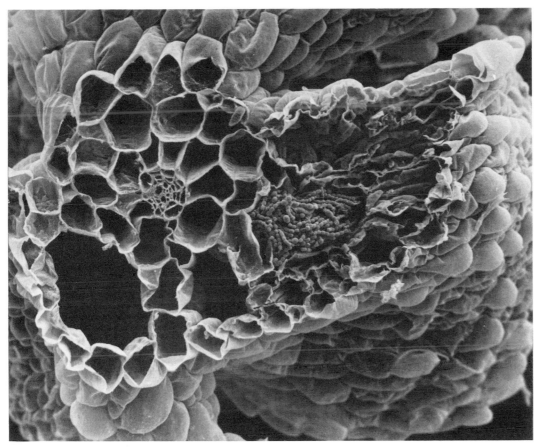

Fig. 3 × 330

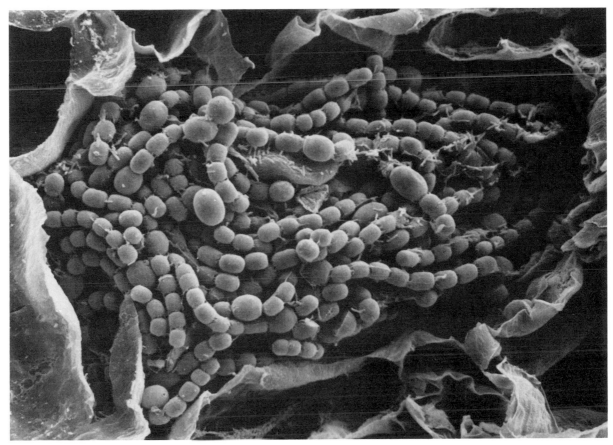

Fig. 4 × 1,550

DUCKWEED

Fig. 1 × 1/2

The duckweed is a tiny flowering plant that floats on the surface of still water, producing dense green mats (Fig. 1). From the bottom of the floating plant dangle roots with bulbous tips (Fig. 2). The body of the plant is a flattened leaf on which tiny flowers are produced consisting of a single petal containing one female (pistil) and two male (stamens) organs (Fig. 3). The stamens produce pollen, which falls on the surface of the pistil (Fig. 4). The sperm from the pollen travel through the pistil to fertilize the eggs, which then develop into a small fruit (Fig. 5), which can germinate to form a new plant.

Fig. 2 × 30

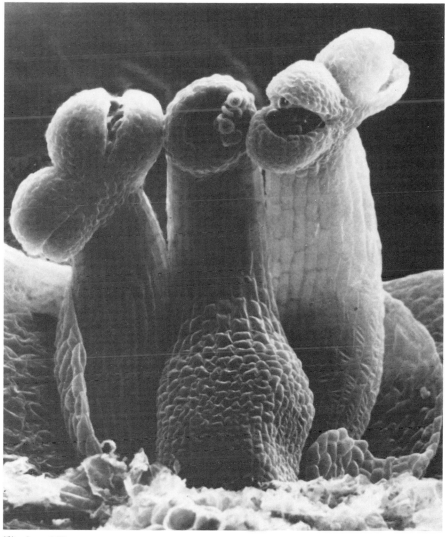

Fig. 3 × 180

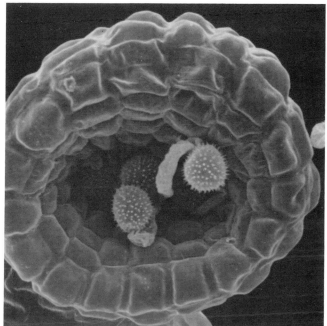

Fig. 4 × 765

Fig. 5 × 150

SHOOT BUDS

Potato

Fig. 1 × 1/4

Fig. 2 × 47

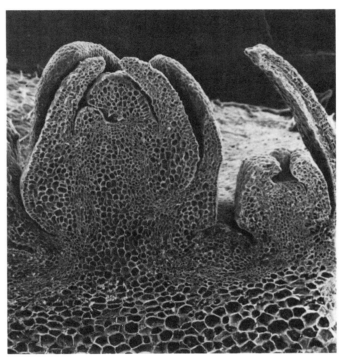

Fig. 3 × 47

Two regions of the higher plants are permanently growing—these are the tip of each stem at which new leaves originate (the shoot apex) and the root tip or root apex. Buds of the underground stem of a potato—the tuber or edible portion (Fig. 1)—provide a good example of shoot apices. Like most shoot tips they are protected by bud scales and closely held surrounding young leaves (Fig. 2). A cut made lengthwise through a shoot tip reveals its beautiful internal structure (Fig. 3). The central portion (the *apical meristem*) consists of cells that will develop into new leaves or occasionally into flowers. A beginning leaf can be seen emerging from the side of the apical meristem as a small bump. Stem veins can be seen forming behind the apical meristem as elongate strands; these will differentiate further to produce cells which transport water and nutrients. Potato buds should never be eaten as they contain a toxic substance.

LEAF FORMATION

Elodea

Fig 1 × 1

Fig. 3 × 134

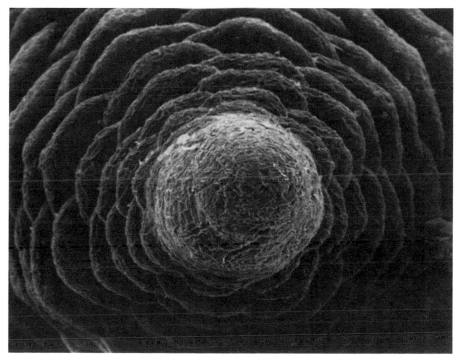

Fig. 2 × 290

The tip of the growing shoot (the shoot apex) of *Elodea* (Fig. 1), an aquatic plant commonly used in aquariums, is unusually long (Fig. 2). This makes it possible to see young leaves in various stages of their development. Many embryonic leaves, which appear as whorls of cells, and immature leaves can be seen in the side view (Fig. 2) and top view (Fig. 3) of the shoot apex. Development begins by enlargement of a small group of cells just under the surface layer of the very tip. These cells then divide and form a small bump on the surface of the shoot apex. As the bump enlarges, cells of the outermost layer begin to differentiate and form the surface layers of the leaf. The inner cells of the bump differentiate to form the leaf veins and other tissue of the leaf.

FLOWER INITIATION
Cocklebur

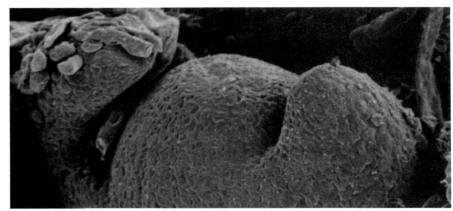

Fig. 1 × 1

Fig. 2 × 385

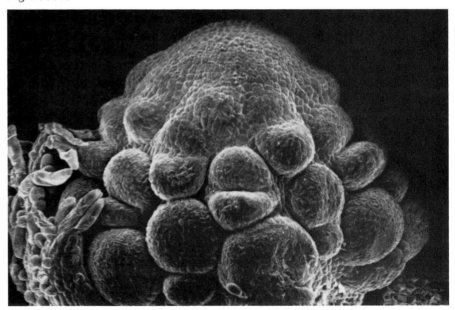

Fig. 3 × 222

The shoot apex of a plant not only produces leaves but also produces flowers. With many plants, of which the cocklebur (Fig. 1) is one example, initiation of flower production is controlled by the length of the day—actually the number of hours of darkness. If the plant is kept in the dark for 8 hours or longer and then returned to the light, a leaf-producing shoot apex (Fig. 2) will produce flower buds instead. Fig. 3 shows the beginning of the formation of a flower bud. Control of budding by light is called *photoperiodism*. A plant that requires a period of darkness to produce flowers is called a short-day plant.

ROOT HAIR

Radish

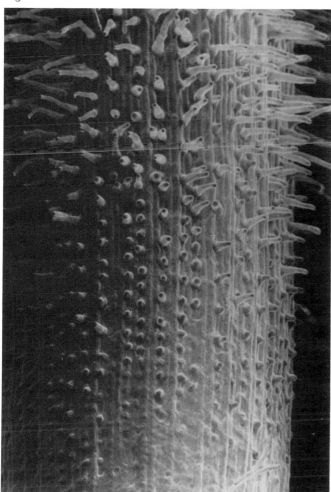

Fig. 1 × 2

The root of a higher plant serves two functions—to anchor the plant in an upright position and to absorb water and minerals from the soil. Absorption is accomplished by *root hairs*, which appear on the root as soon as the seed germinates (Fig. 1). Root hairs are slender, tubular extensions from the cells on the root surface (Fig. 2). They serve to increase the surface area of each surface cell so that the cell can absorb water more rapidly. Root hairs are located in a region immediately behind the elongate part of each root fiber (Fig. 3). Individual root hairs are short-lived and wither as the root elongates but new hairs form continually just behind the portion of the root that is elongating.

Fig. 2 × 95

Fig. 3 × 40

ROOT

Ginseng

Fig. 2 × 1

Fig. 1 × 1/4

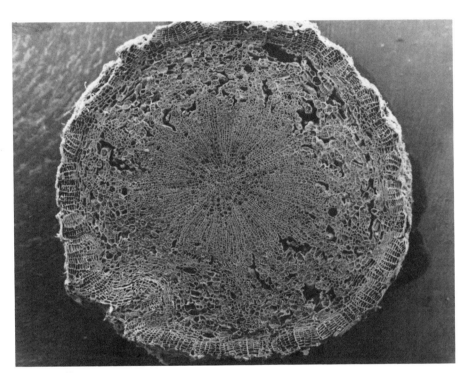

Fig. 3 × 43

The ginseng root in some way resembles a human and for that reason it has been considered in the Orient to have mystical and health-restoring properties. The American ginseng is a fleshy-rooted herb (Fig. 1) that grows in the shady hardwood forests of northeastern America. The root is crowned with an underground stem called the neck (Fig. 1), which grows upward. The neck is extremely sensitive to cold and would be damaged during the winter were it not for the fact that as the neck grows, the root shrinks, pulling the base of the neck downward. Examination of a cross-section of a branched root shows that the root is protected by several folded layers of cells (Fig. 2), which gives the root a wrinkled surface. The inner portion of the root (Fig. 3) is similar to that of higher plants, consisting of a central core of water-transporting cells; between the arms are groups of large cells responsible for transport of nutrients (Fig. 4).

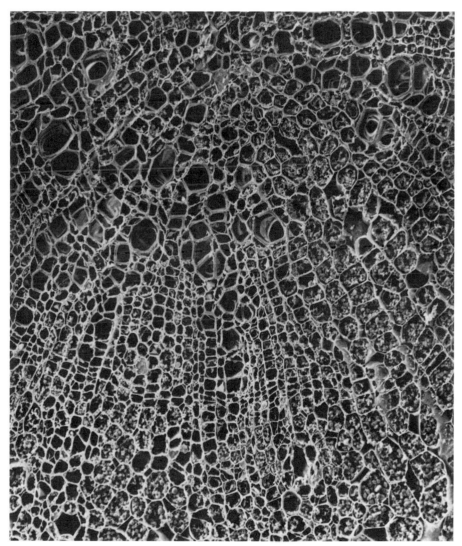

Fig. 4 × 250

STEM

A broad-leaved plant: Tomato

Fig. 1 × 1/4

The stem of all higher plants is the link between the root and the leaves. It serves to transport water upward through *xylem* cells and downward through *phloem* cells and it supports the leaves. Some stems are also storage organs, such as the underground stem of the potato, while others, such as those of the cactus and other desert plants, store water and carry out photosynthesis. Plant stems are either woody or soft (herbaceous). The tomato plant (Fig. 1) has a soft green herbaceous stem in which there are circular bundles of xylem and phloem cells—the large and small cells in the ring (Fig. 2)—which separate the outer part of the stem from the central pith, which serves to store food. The entire stem is enclosed in a skin-like layer consisting of cells having many external hairs.

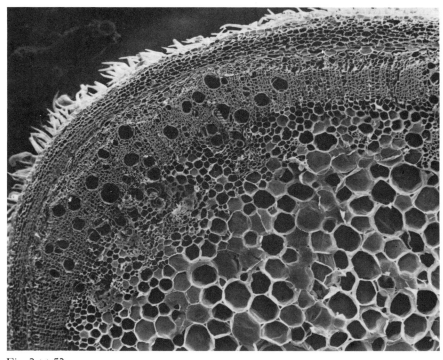

Fig. 2 × 53

STEM

Grasses: Corn

Fig. 1 × 1/10

Corn (Fig. 1), like all grasses, also has a herbaceous stem. Its structure differs from that of the tomato stem, seen on the preceding page, in that the bundles of xylem and phloem cells are not organized in a ring but are scattered throughout the other cells of the stem (Fig. 2); the bundles are more numerous in the outer part of the stem though (Fig. 3). This arrangement is characteristic of the grasses. Each bundle is enclosed in a sheath of somewhat woody cells which contributes to the strength of the stem. The central cells of the stem are used for storage of nutrients. In some grassy stems, for example those of wheat and bamboo, most of these central cells disintegrate, leaving a hollow core.

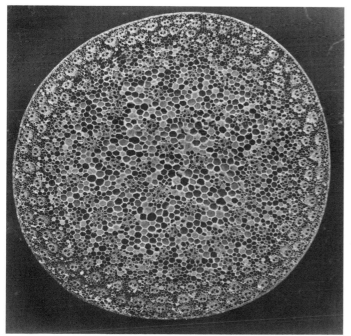

Fig. 2 × 22

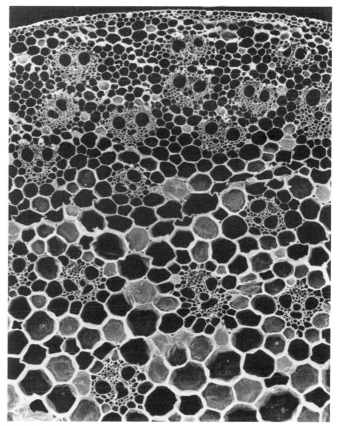

Fig. 3 × 87

NEEDLES AND WOOD
Pine

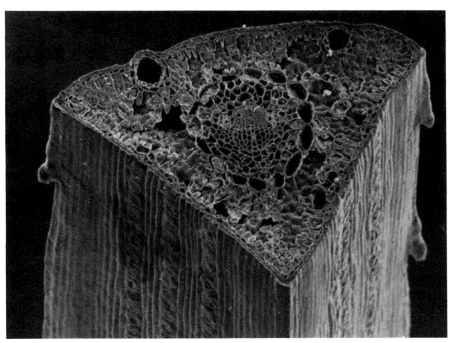

Fig. 1 × 1/2

Fig. 2 × 134

Pine trees have long needle-like leaves arranged in a bundle on a short spur shoot (Fig. 1). The needles have a triangular cross-section (Fig. 2) and contain microscopic pores for releasing water vapor and for secreting a resin used for wound healing. Also there is a central vein through which water and nutrients flow. Pine wood, like the wood of most conifers, contains darkened patterns produced by several different cell types. One of these cell types is the *tracheid* (open cells at the upper edge of Fig. 4), which is used primarily for transporting water. The tracheids form in a layer surrounding the trunk just under the bark. This happens in the spring and early summer, during which time the cells are large and the cell wall is very thin. In autumn the tracheids become smaller but their cell walls become very woody and thickened; this produces the annual rings seen in a cross-section of wood (Fig. 3). A block showing the three faces of pine wood is shown in Fig. 4. The interweave of different cell types, which gives different patterns of wood grain in different cut faces, is clearly displayed. The open structure of pine wood accounts for its lightness and is characteristic of softwood.

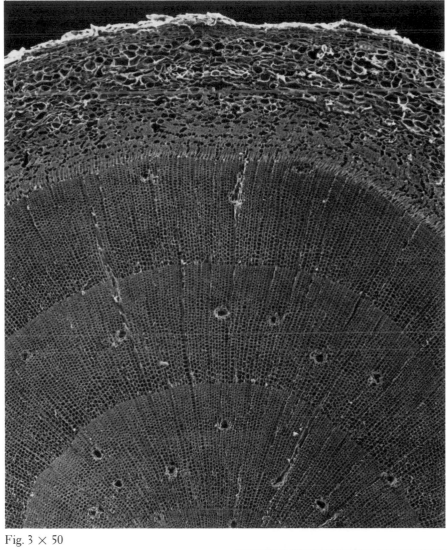

Fig. 3 × 50

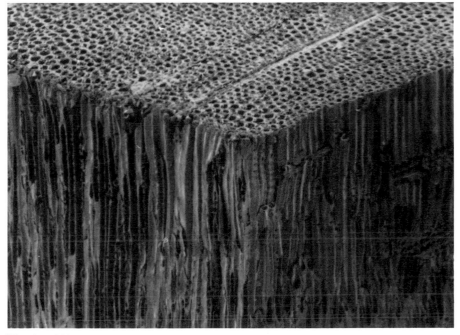

Fig. 4 × 108

WOODY STEM

Oak

Fig. 1 × 1/2

Fig. 2 × 59

The wood of broad-leaved trees such as oak (Fig. 1) is more complex than that of the conifers, such as the pine, seen earlier. Most noticeable are very large cells called *vessels* (Fig. 2); these cells are shorter than the tracheid cells of the pine but have a much larger diameter. As the plant grows, the end walls of the vessel cells break down forming a continuous tube through the stem. The walls of the tube are perforated with numerous openings forming interesting patterns (Fig. 3). In older wood some outgrowths of adjacent cells penetrate the tube (Fig. 4). A woody stem can be cut in different planes; three of these—cross-section, vertically through the center of the stem, and vertically but perpendicular to the radius of the stem—enable one to view how different types of wood cells are interwoven (Fig. 2).

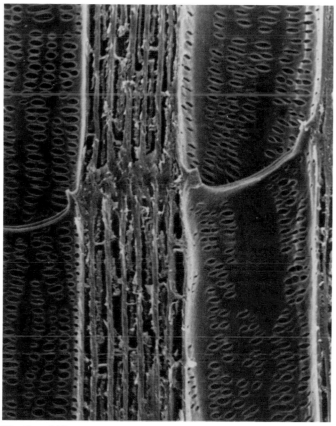

Fig. 3 × 560

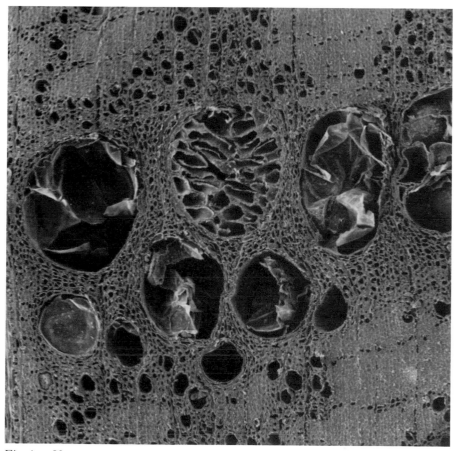

Fig. 4 × 90

LEAF

Broad-leaved plant: Geranium

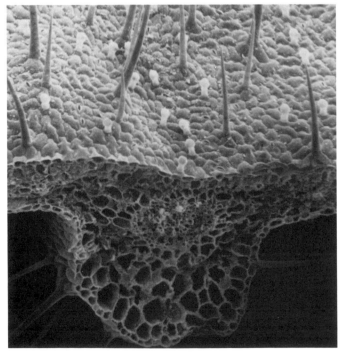

Fig. 1 × 1/2

The main function of a leaf is to carry out photosynthesis—that is, to synthesize glucose from carbon dioxide and water. The glucose is then transformed chemically into all of the organic materials needed by the plant. The geranium, a typical leaf (Fig. 1) is enclosed in a layer of cells (*epidermis*) on both the upper and lower surfaces of the leaf. The epidermis on the lower side of the leaf has numerous openings (*stomata*) through which water can leave the plant and carbon dioxide can enter the plant. Hairs and secretory glands, which often produce a waxy coating that protects the plant from intense sunlight and drying, are found on both surfaces (Figs. 2, 3). The internal tissue of the leaves of most broad-leaved plants consists of two layers of cells—a *palisade layer* on the top (upper layer in Fig. 3) and a *sponge layer* on the bottom. The veins consist of organized bundles of *xylem* and *phloem* cells and are located between these layers (Fig. 2). Some of the larger veins often protrude from the under surface of the leaf.

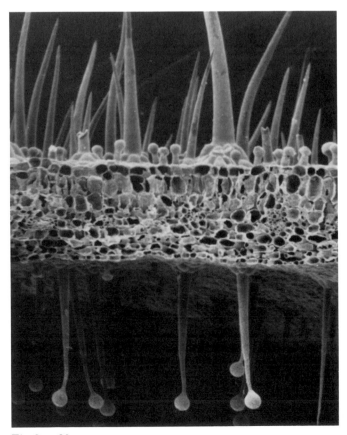

Fig. 2 × 86

Fig. 3 × 90

LEAF

Narrow-leaved plant: Corn

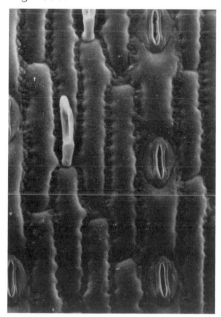

Fig. 1 × 70

Fig. 2 × 425

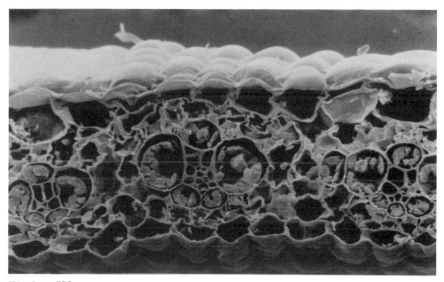

Fig. 3 × 523

The leaf of most narrow-leaved plants, such as the grasses, consists of a narrow blade and a sheath enclosing the stem. The outer layer (Fig. 1) consists of more cell types than that of the leaf of a broad-leaved plant; the principal cell is a narrow elongate cell. The openings, or *stomata*, on the under side of the leaf are fringed with several cell types— triangular cells and guard cells—which, by engorging or releasing water, change their shapes, and enable the stomata to be either open or closed. In the figure the stomata are all nearly closed. The long rectangularly shaped cells in Fig. 2 are the most common cell type on the surface of leaves. The interior of this leaf is not organized into a palisade and spongy layer as in the leaves of the broad-leaved plant and the bundles of *xylem* (water-transporting cells) and *phloem* (nutrient-transporting cells) are enclosed in a sheath called the *bundle sheath* (Fig. 3). In some tropical grasses, the bundle sheath has a specialized function which enables the leaf to take in carbon dioxide more efficiently without keeping the stomata wide open. This is necessary in the dryer tropical regions to avoid excessive drying of the leaf by water loss through the stomata.

LEAF
Rose and Bean

Fig. 1 × 1/2

An interesting view of the inner structure of a leaf can be obtained by peeling off the leaf epidermis. This has been done with a rose leaf (Fig. 1). When the epidermis of the upper layer of a rose leaf is removed, the palisade layer can be seen as a densely packed layer of cells of nearly uniform size (Fig. 2). The depressions are the air spaces directly behind the stomata. In these air spaces one can see the elongate form of the palisade cells; these cells contain the photosynthetic organs and provide a long path for light rays so that a great deal of the light energy can be captured by each cell. When the lower layer of the leaf is removed, the larger, irregularly shaped sponge cells can be seen (Fig. 3). Note the large air spaces between the cells. Fig. 4 shows the result of removing the lower epidermis from a bean leaf in such a way that the peeling has passed through every layer of the leaf. Here the complete anatomy of the leaf is very well shown. From top to bottom, the layers include the underside of the lower epidermis, the sponge layer, a leaf vein, the palisade layer, and the upper epidermis.

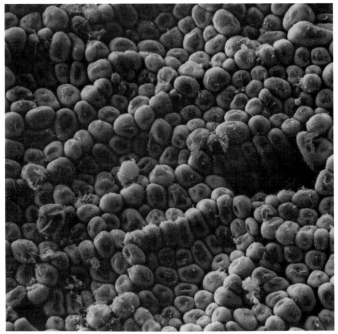

Fig. 2 × 755

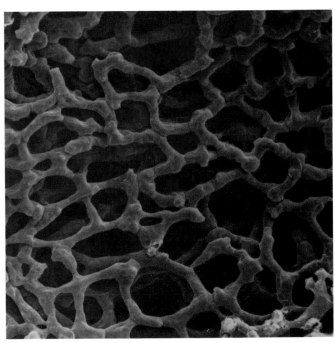

Fig. 3 × 470

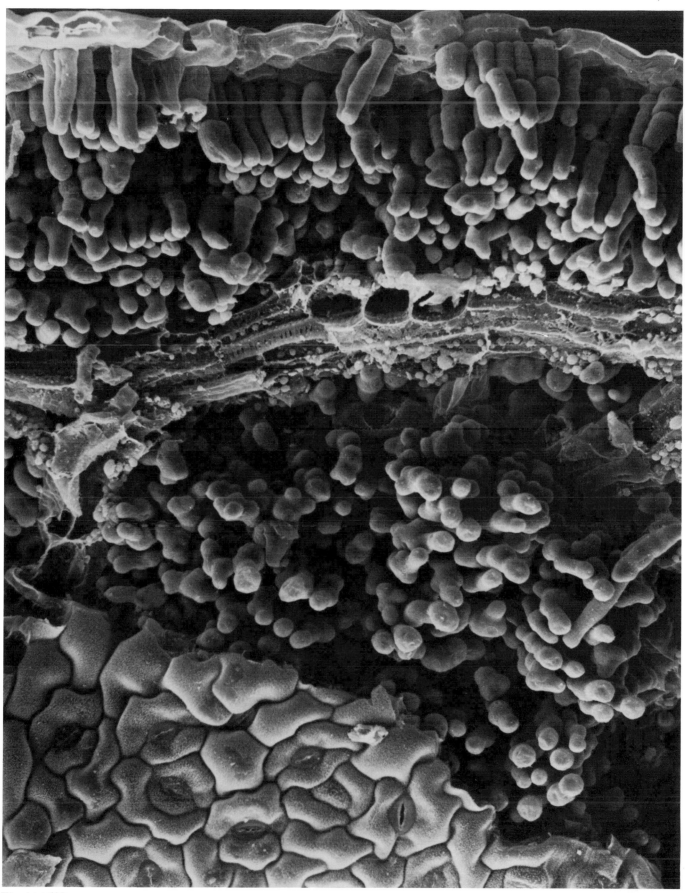

Fig. 4 × 600

LEAF VEINS

Eucalyptus

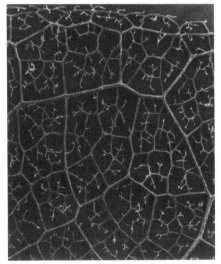

Fig. 1 × 1/3

Photosynthesis, the production of glucose from carbon dioxide and water, occurs in green leaves. The leaf veins consist of two types of cells—the *xylem* and *phloem*. The xylem brings the water to the photosynthetic apparatus of the leaf and the phloem carries the newly-synthesized glucose from the leaf to the stem. Veins can be examined by dissolving away all of the soft tissue of the leaf, leaving behind only the vein system. Veins are organized in one of two ways. In narrow-leaved plants (*monocots*) the veins are generally parallel to one another. In the broad-leaved plants (*dicots*) shown here the veins form a network (Figs. 1, 2). In the dicot leaves the network consists of one or more prominent veins from which smaller veins branch and join with other veins (Fig. 1). The finest veins terminate in clusters of open-ended veins for delivery to and pick-up from nearby cells (Fig. 3).

Fig. 2 × 7.5

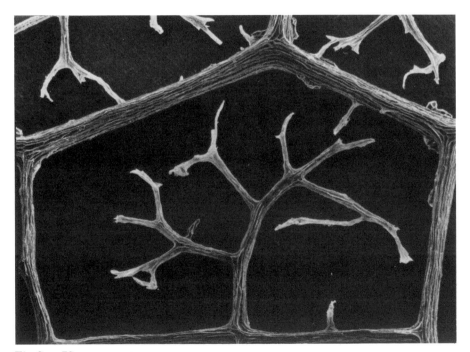

Fig. 3 × 70

PHLOEM AND SIEVE TUBES

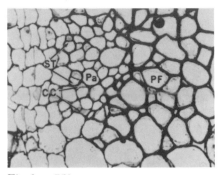

Fig. 1 × 750

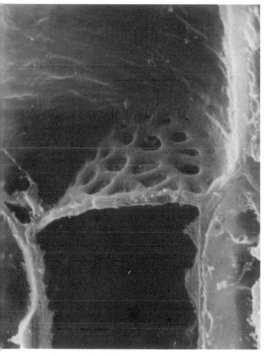

Fig. 2 × 3,850

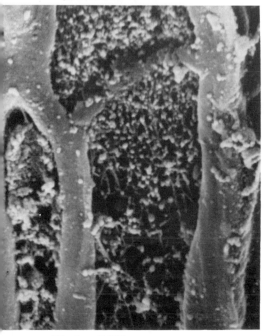

Fig. 3 × 7,000

Phloem is a complex tissue consisting of several types of cells. One type transports the sugar made in the photosynthetic apparatus of the leaves to the rest of the plant. Another type of cell, having thicker walls, which are often woody, provides strength to the system. Phloem tissue is not usually spread throughout the vein or stem but is organized into cell clusters known as *phloem bundles*. A cross-section of a phloem bundle is shown in Fig. 1. The nutrient transport cells are called *sieve tubes* and are labeled ST in the figure. These are elongate tubular cells at the end of which is a disc having numerous pores (Fig. 2); this structure is called a *sieve plate*. Nutrients can flow in either direction in the sieve elements, carrying sugars made in the leaves to the stems and roots and organic chemicals made in various parts of the plant to the leaves and flowers. This flow pattern contrasts with that of the *xylem*, or water-transporting tubes, through which water flows only from the roots to the leaves. The inner wall of a sieve tube is lined with numerous small clusters of a fibrous protein called *phloem protein*. If a leaf vein or stem is torn, this protein is released from the wall of the sieve tube. The protein molecules aggregate to form a pasty substance that clogs the pores of the nearest sieve plate (Fig. 3). This prevents nutrients from draining from the plant. In time a callose is formed around the sieve pore to aid in plugging the site of the break.

LEAF APPENDAGES
Trichomes of marijuana and tobacco

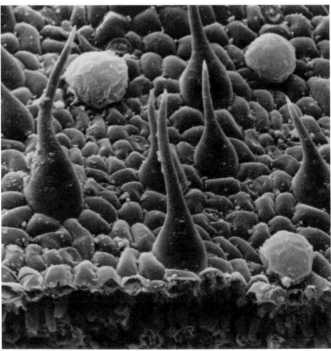

Fig. 1 × 55

Trichomes are small secretory elements on the surface of most leaves, having a variety of shapes (Fig. 1). Two or more shapes are often found in the same region of the surface such as on the leaves of marijuana (Fig. 2) and tobacco (Fig. 3). Trichomes secrete a variety of substances, which by virtue of their smell or taste, repel many plant-eating insects. Some of these substances such as tetrahydrocannabinol (THC) and nicotine are stimulating rather than repellent to humans so that humans inhale the smoke of these leaves in order to have pleasant sensations.

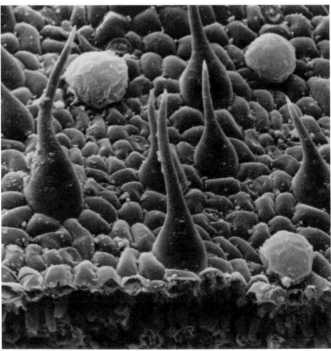

Fig. 2 × 326

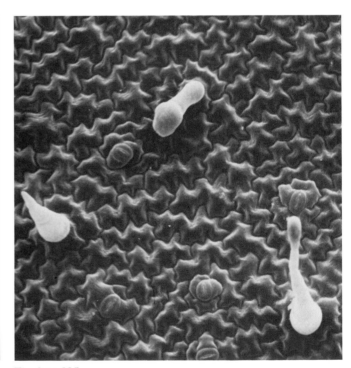

Fig. 3 × 225

MODIFIED LEAF

Venus Flytrap

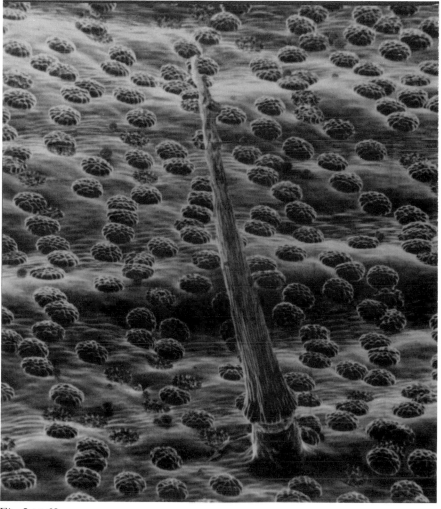

Fig. 1 × 1/2

The Venus flytrap (Fig. 1) is a well-known insect-catching plant, native to swamps of the eastern United States. The plant traps and digests insects to obtain nitrogen and minerals that are lacking in the swamp. The leaves consist of two units which form a cage that can be opened and closed. When closed, a comb of bristles interlocks and prevents the trapped insect from escaping. Sensitive hairs on the surface of the leaf respond to being touched by closing the cage. The trapped insect stimulates many hairs which causes glands on the leaf surface (Fig. 2) to secrete digestive enzymes.

Fig. 2 × 60

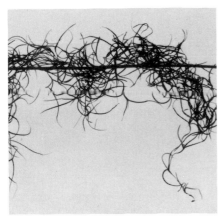

Fig. 1 × 1/2

AIR PLANT

Spanish moss

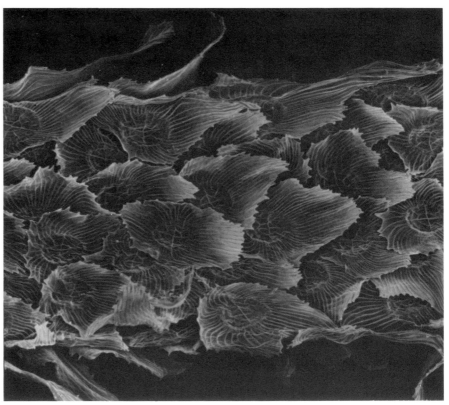

Fig. 2 × 102

Spanish moss is neither from Spain nor is it a moss. In contrast with its appearance, it is a relative of the pineapple. The plants are called *epiphytes*, which means that they grow supported on other plants. Spanish moss is usually found draped from the branches of trees in humid tropical regions and occasionally on telephone wires (Fig. 1). This habit of growth has given it the name air plant. Spanish moss has no roots and gathers all of its water from rain. It obtains nutrients from liquids washed from its supporting plants by the rain. The stems and leaves are covered with flattened hairs (Fig. 2) which create a small space in which rain water is trapped, rather than running off immediately. A cluster of four cells forms an elegant array (Fig. 3) and is responsible for absorbing water and nutrients.

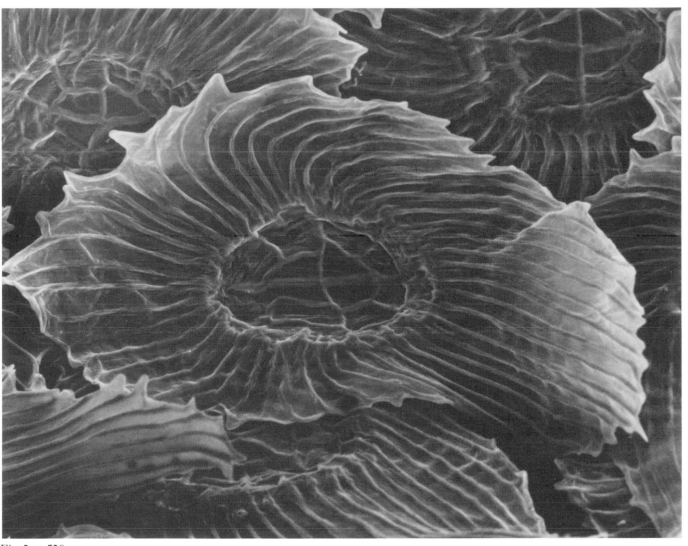

Fig. 3 × 538

FLOWER FORMATION
Corn

Corn is usually not thought of as a flowering plant but has two types of flowers—male flowers grouped at the top of the plant forming a tassel and female flowers that ultimately form the fruit, namely, the ear of corn. The tassel forms initially from small projections on the side of the shoot tip (Fig. 1). Fig. 2 is an enlargement of a portion of the tip shown in Fig. 1; the flowers develop from the knobs. In time the knobs differentiate; Fig. 3 shows the nearly mature male flower with the three developing stamens when they first appear as distinct structures. The ear begins as a sharply pointed tip (Fig. 4). Each bump is a site at which one kernel will form but the bumps first develop into female flowers. A high magnification of a more developed portion of the tip shows the components of the flower-to-be. Fig. 6 shows a more advanced stage than the units in Fig. 5. One of these projections will be the pistil and will extend to become corn silk.

Fig. 1 × 82

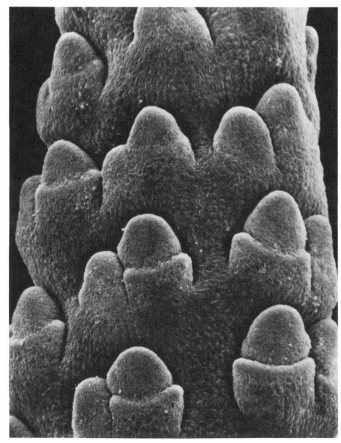

Fig. 2 × 200

Fig. 3 × 80

Fig. 4 × 27

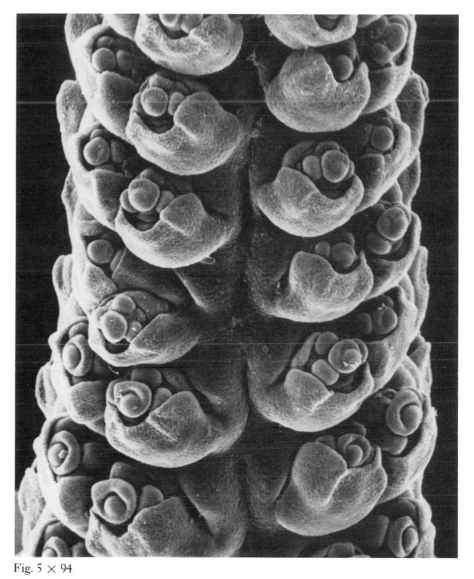

Fig. 5 × 94

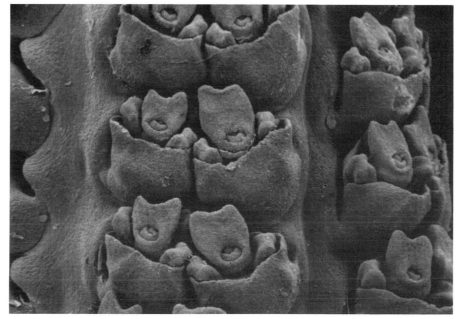

Fig. 6 × 72

FLOWERS

Mustard, Rose, and Geranium

Fig. 1 × 2

The flowers of broad-leaved plants have four components—*sepals* (which enclose the bud), *petals*, pollen-producing (male) *stigmas*, and seed-producing (female) *pistils*. In a mustard flower (Figs. 1, 2) one can see 4 sepals, 4 petals, 6 stamens, and 2 pistils. The most noticeable difference between flowers is their shape and color. Mustard is yellow, the rose (Fig. 5) can be many shades of red, yellow, or white, and the geranium (Fig. 6) is red or white. Electron micrographs of the petals of mustard (Fig. 3), the rose (Fig. 4), and the geranium (Fig. 7) show that the textures of the petals differ markedly.

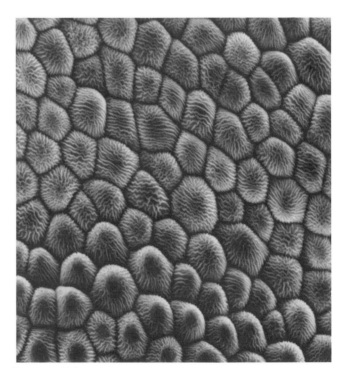

Fig. 3 × 815

Fig. 2 × 12

62

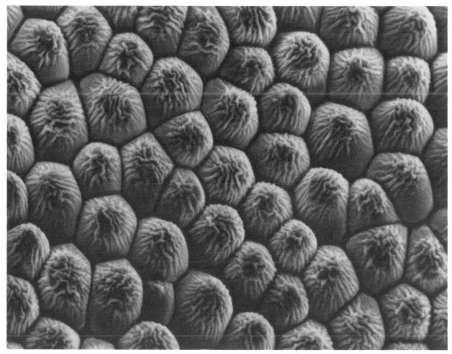

Fig. 4 × 725

Fig. 5 × 1/2

Fig. 6 × 1/2

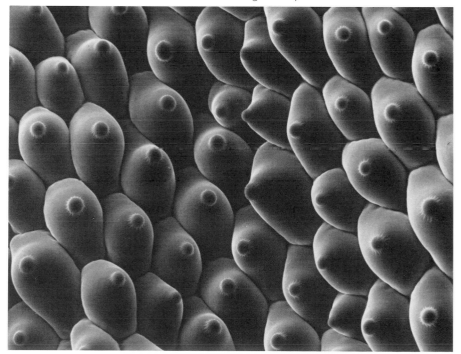

Fig. 7 × 315

STAMEN AND POLLEN GRAINS

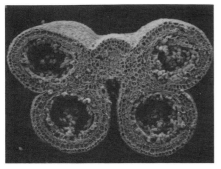

Fig. 1 × 30

The *stamen*, or male organ of a flower, has two components—a thin stalk arising from within and near the base of the flower and a pollen-producing organ, the *anther*, which is located at the end of the stalk. Most flowers have several stamens. When examined in cross-section, the anther can be seen to contain two to four chambers (Fig. 1). Pollen grains form in the chambers and are released when the chambers mature and burst. When mature, each anther is usually covered with a thick layer of pollen dust. Pollen grains differ widely in their size and shapes from one species of plant to another. These differences are evident in the micrographs shown of pollen brushed from flowers of ragweed (Fig. 2), African violet (Fig. 3), geranium (Fig. 4), duckweed (Fig. 5), lemon (Fig. 6), and snapdragon (Fig. 7). In order for seeds to form, a pollen grain must fall on the tip of the female organ, the pistil, germinate, and make its way to the base of the pistil where the eggs are located. Contact with the pistil can occur by growth of the pistil to the anther, as explained on page 66, or the pollen can be transported by wind or insects. Pollen, especially that of ragweed, grasses, and some trees, is the cause of many allergies in man. Some pollen grains are used in cooking; the pollen of the crocus is known as saffron.

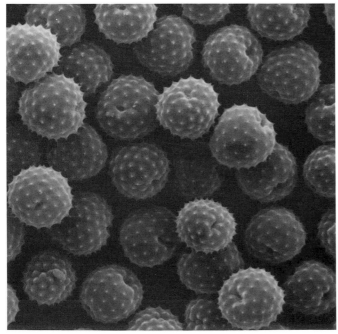

Fig. 2 × 860

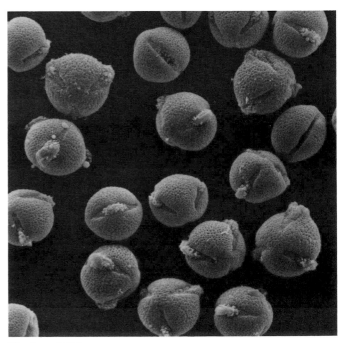

Fig. 3 × 1,010

Fig. 4 × 650

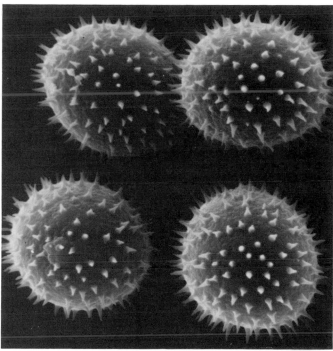

Fig. 5 × 2,690

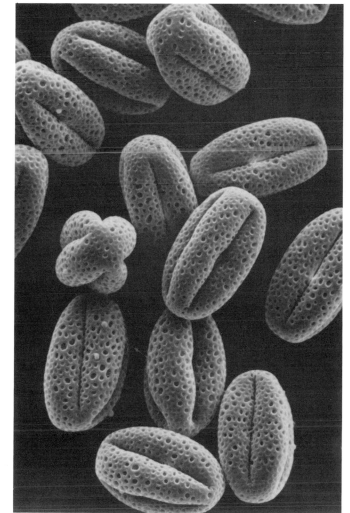

Fig. 6 × 950

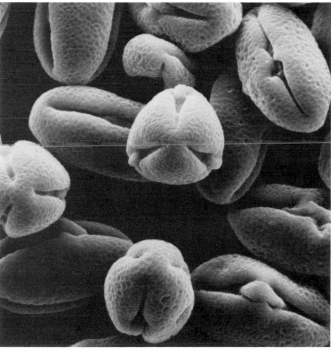

Fig. 7 × 1,550

STIGMA AND POLLINATION

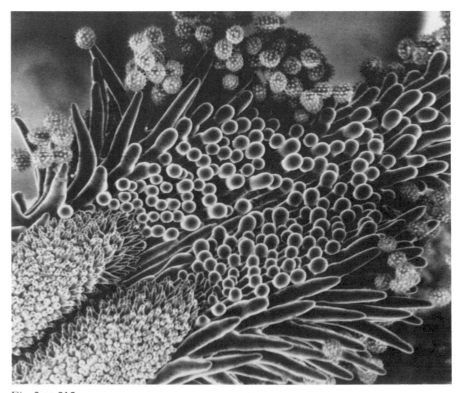

Fig. 1 × 23

Pollination, or the transfer of pollen from the anther on which it formed to the stigma of the same or another flower, occurs in a variety of ways—for example, by wind, insects, and by movement of various parts of the flower. An example of the latter can be seen in the bidens, a composite flower, one of whose florets is shown in Fig. 1. The anthers are joined together to form a ring about the tip of the pistil. As the stigma at the tip of the pistil lengthens, it pushes through this ring and ruptures the pollen sacs. The pollen then falls directly onto the stigmas. A close-up view of the stigma after contact with an anther is shown in Fig. 2; the spiked pollen grains can be seen at the top and to the right in the figure. Once a pollen grain lands on the stigma surface, a hollow tube forms within the pistil—from the stigma to the base of the pistil where the ovary is located. The pollen grain responsible for forming the tube moves down toward the ovary, which contains the eggs. As this is occurring, certain cells in the pollen grain differentiate and become sperm cells, which fertilize the eggs, thereby initiating formation of the seed.

Fig. 2 × 215

DISPERSION OF SEEDS

Dandelion

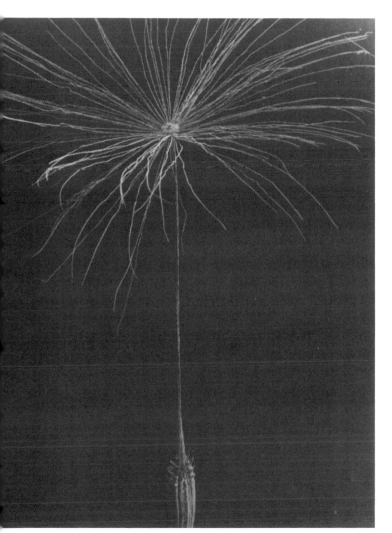

Fig. 1 × 1

It is advantageous for plants to disperse their seeds so that new seedlings crowd neither the parent plant nor one another. Seeds can be carried by water, animals, and wind. The dandelion seed is dispersed by wind. This flower is a composite flower, which, when gone to seed, produces a familiar spherical cluster of plumed seeds (Fig. 1). Fig. 2 shows one of the plumed seeds, which is the fruit of one of the florets of the composite flower. An enlargement of the center of the plume is shown in Fig. 3. When mature, the plume is torn from the cluster by the wind and often carries the seed several hundred yards.

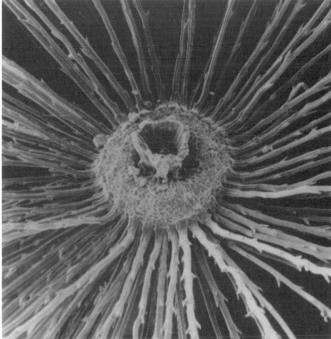

Fig. 3 × 90

Fig. 2 × 10

SEED GERMINATION
Corn and Barley

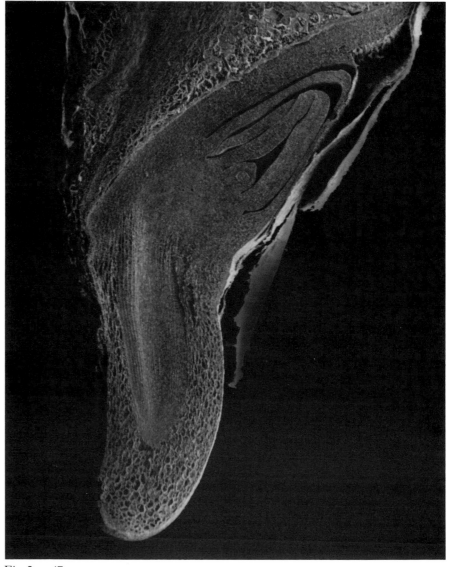

Fig. 1 × 2

Seeds consist of an embryonic plant and food storage tissue. The latter part may be a single unit as in barley (Fig. 1) or two units as in a peanut. A single-unit seed is called a *monocot*; one with two units is a *dicot*. A slice through a barley seed shows that the embryo is already differentiated. The seed is enclosed in a sheath that protects the young leaf and sheet (upper right) during germination. The lower part of the embryo (projecting downward) is a primitive root.

Fig. 2 × 47

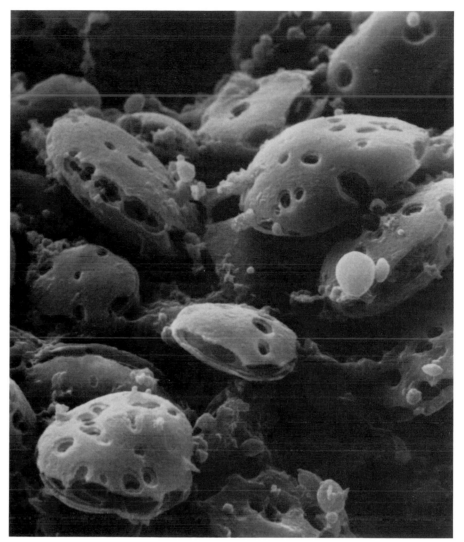

Fig. 3 × 3,135

The initial growth of the seed, which generally occurs in the absence of light, utilizes nutrients formed in the storage tissue at an earlier time while the seed was developing. Various carbohydrates are used as a storage substance but in barley the main storage substance is starch. During germination of a barley seed, a plant hormone, called *gibberellin*, activates an enzyme that converts the starch to sugars that the embryo can utilize as food. Interestingly, this enzyme, which is called *amylase*, is very closely related to the enzyme in animal saliva whose main purpose is to convert various vegetable starches to sugars before the starch reaches the stomach. In barley, digestion of the starch begins in well-separated locations, creating numerous microscopic pits; many of these pits can be seen in Fig. 3. In time, progressive solubilization of successive layers of starch along the margin of the pit causes the pit to enlarge. Ultimately all of the starch is dissolved and the food reserves are gone. However, by this time the young seedling has emerged from the ground and synthesized chlorophyll. The chlorophyll enables the plant to utilize the energy of sunlight to make its own glucose by photosynthesis.

PROTOZOA

Ciliates

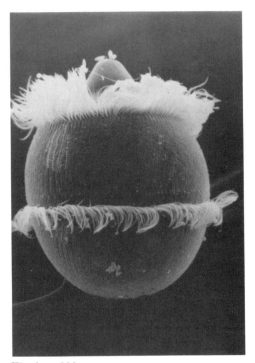

Fig. 1 × 800

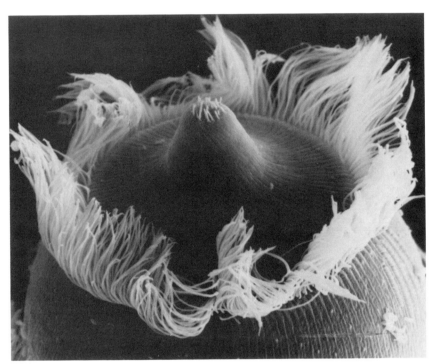

Fig. 2 × 2,285

Most stationary fresh water abounds with protozoa, one-celled organisms considered to be the most primitive animal cells. Protozoa are generally motile, they can often ingest solid food particles, they store their genetic material in a nucleus (in bacteria genetic material is present throughout the cell), and they lack the rigid cell wall of plant cells. There are 60,000 species of protozoa, which have a variety of properties and forms and which range in size from microscopic to a few millimeters; some can be seen by a person with sharp eyes. Protozoa having a large number of tiny hairs (*cilia*) used for locomotion and food gathering are called ciliates. The ciliate *Didinium* is shown in Figs. 1 and 2. This is a spherical cell on which the cilia form two bands encircling the organism. *Didinium's* "mouth" is at the end of a small projection, shown in Fig. 2. Three ciliates that are uniformly covered with cilia are shown in Figs. 3–5. *Paramecium* (Fig. 3) has a groove that serves as a mouth. *Spirostomum* (Fig. 4) is a giant among ciliates and may be as much as 3 mm long. Its cilia are arranged in spiral bands and it has one thicker band that can be seen in the photo. *Blepharisma* (Fig. 5) is a pink ciliate whose cilia are in long rows. It has a region free of cilia that has a special function.

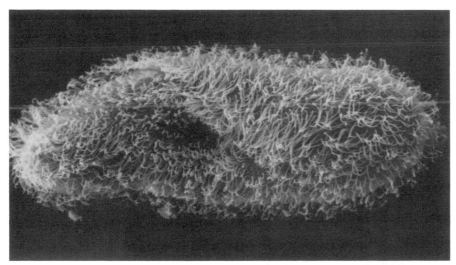

Fig. 3 × 1,210

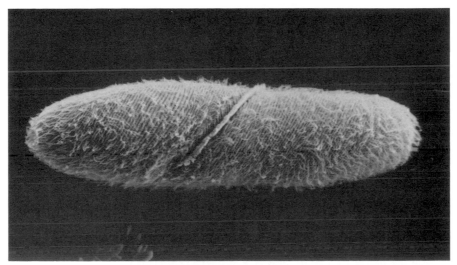

Fig. 4 × 335

Fig. 5 × 640

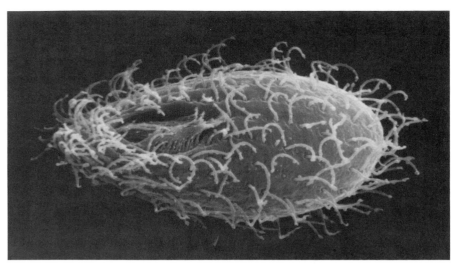

Fig. 6 × 3,380

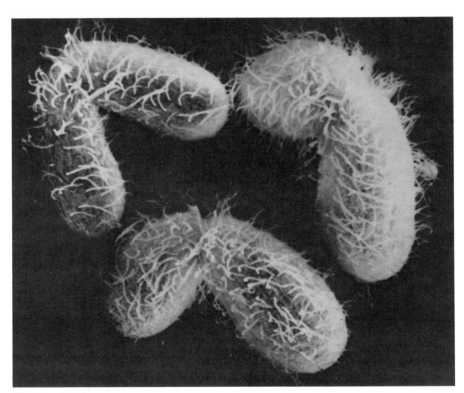

Fig. 7 × 1,790

Tetrahymena pyriformis (Fig. 6) is a common ciliate, about a half millimeter long, widely studied in research laboratories. Its cilia cover the entire body. The shallow depression seen in the upper figure is the oral cavity or mouth. The cilia have a special arrangement within the mouth in that they are organized into three bands, one of which is visible, which, by their wave-like beating, direct food into the mouth. *Tetrahymena* reproduce both asexually by simple division and sexually by mating. Fig. 7 shows three pairs of mating cells. The cells attach on their bottom surface and genetic material is exchanged. Each cell acts as both male and female and donates genetic material to the other. After an internal rearrangement the cells separate and each divides.

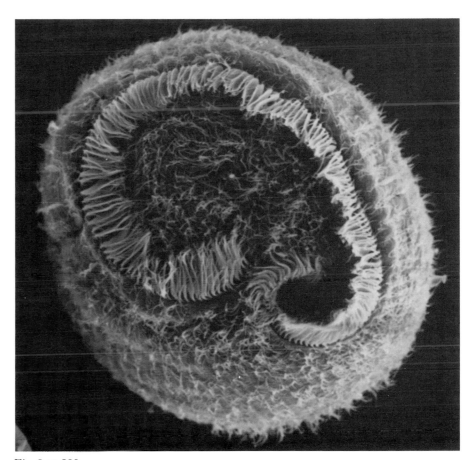

Fig. 8 × 900

Stentor (Fig. 8) is a long conical protozoan. It is one of the few protozoans that is not primarily free-swimming. It is often found attached to solid surfaces such as rocks and leaves by the terminus of its flexible narrow region. Its uppermost wide part is a hollow cup, the opening being its oral cavity. The figure shows a view into this cup-like portion. Both the exterior and interior regions of *Stentor* are heavily ciliated. Some of the cilia within the oral cavity are fused together to form membranous bands. These bands wave back and forth in such a manner that water flows into the mouth. Food particles caught in the water currents are thereby carried into the mouth. Waste products are excreted through a small opening in the tail of the cell. *Stentor*, like most protozoa, reproduces both asexually and sexually. In the asexual mode, which is the more common, a mature *Stentor* splits in half to yield two daughter cells. In the sexual mode two *Stentor* come in contact, exchange genetic material, separate, and then each divides, as in the case of *Tetrahymena*.

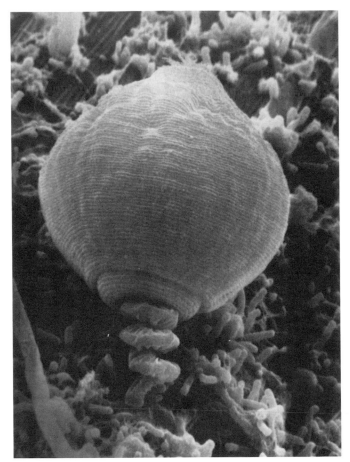

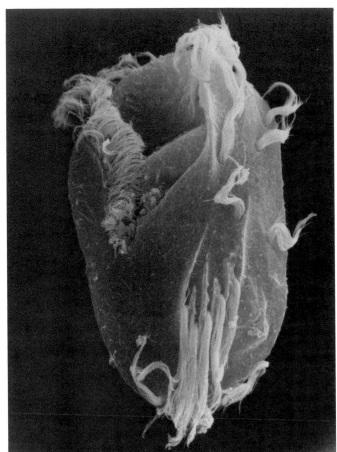

Fig. 9 × 2,960

Fig. 10 × 760

Vorticella (Fig. 9) is the comic of the protozoan world. It consists of a slightly pointed spherical body connected to a long flexible stalk. The stalk is generally attached to a solid surface, such as a leaf or a piece of wood. The stalk is capable of coiling and uncoiling, which it does repeatedly. The result is that when observed in a light microscope, *Vorticella* seems to be bouncing up and down quite rapidly. In the photo shown the stalk is coiled. *Vorticella* has only a small ring of cilia at the top of the spherical body, surrounding the mouth. These are hidden from view in the photograph. The ciliate *Euplotes* (Fig. 10) has a body that has distinct top and bottom surfaces. Each surface has highly modified cilia. Adjacent to the oral cavity is a row of cilia that form a thin membrane which is used both for locomotion and for directing water currents toward the mouth. In this way food particles are swept into the mouth. The lower part of the figure shows several clusters of cilia that are fused to form bristle-like structures. These produce a strong beating motion that enables the cell to move in circles.

PROTOZOA

Radiolaria

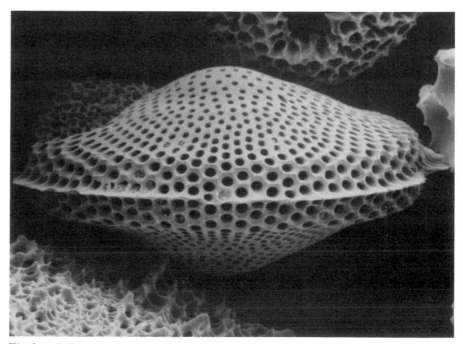

Fig. 1 × 50

Radiolaria are microorganisms consisting of an easily deformable, almost shapeless body, encased in an elegantly decorated hard skeleton (Fig. 1) which is made of a substance identical in composition to sand. The skeletons, which have a variety of lovely shapes (Figs. 2–9), contain many pores through which portions of the body can move in and out when seeking food. Radiolarians are found exclusively in the oceans where they are exceedingly numerous. When a radiolarian dies, the soft part of its body disintegrates but its skeleton sinks to the bottom of the ocean. The sand-like shells are extremely resistant to chemical change and have lasted for hundreds of millions of years. These organisms are so numerous and the accumulation has gone on for so long that the ocean floor is covered with a deep layer of radiolarian shells mixed with other fine particles, a mixture known as *radiolarian ooze*. In deep parts of the ocean, the water pressure is so great that in time the ooze is compressed into a sandy rock. Since many parts of the sea floor have been uplifted as the earth has deformed over millions of years, radiolarians have been found as fossils in rock on the surface of the earth. Many modern radiolarians are identical in shape to those that lived millions of years ago.

Fig. 2 × 556

Fig. 3 × 810

Fig. 4 × 895

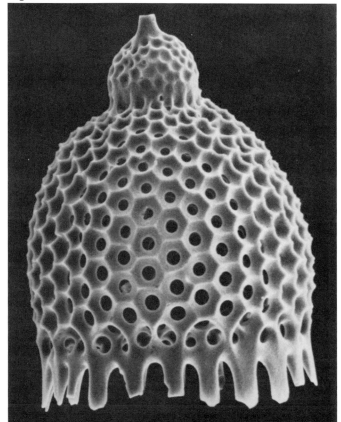

Fig. 5 × 695

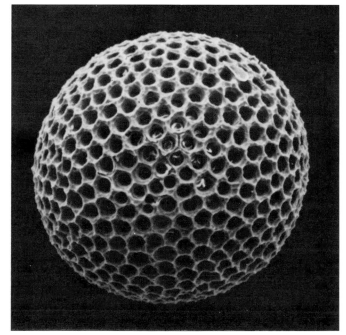

Fig. 6 × 114

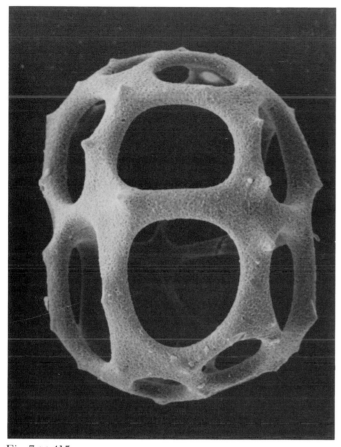

Fig. 7 × 415

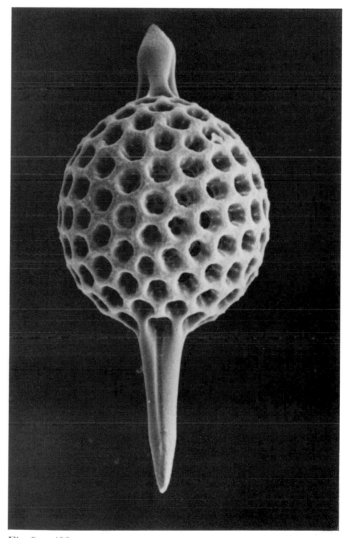

Fig. 8 × 480

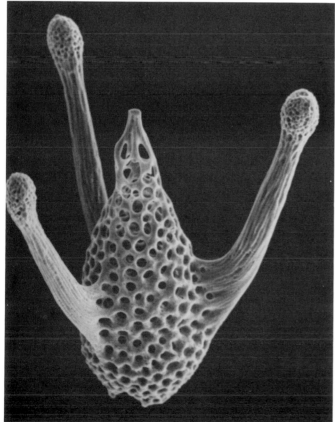

Fig. 9 × 640

77

PROTOZOA

Foraminifera

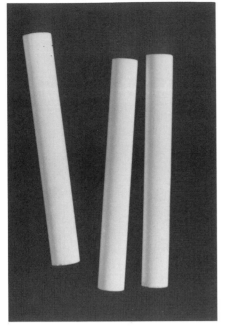

Fig. 1 × 3/4

The *Foraminifera* are marine organisms similar to the radiolarians just described except that their shells are made of a calcium compound instead of being sandlike. The shells comprise a major portion of limestone rock and chalk (Fig. 1). Their intricate and artfully designed shells are, like radiolarians, also covered with pores (Fig. 2) through which tiny fingers of the soft body of the animal extend when seeking food; each shell usually contains a large pore referred to as a window (Fig. 2). The shells also commonly contain chambers of various sizes (Figs. 3, 4). When the organism dies, the shell falls to the ocean bottom, forming a sediment known as *foraminiferan ooze*, which over a period of millions of years has been compressed to limestone. Whereas radiolarian ooze has been found in the deepest ocean trenches, foraminiferan ooze is rarely found below 15,000 feet, for at greater depths the increased carbon dioxide content of water causes the shells to dissolve.

Fig. 2 × 134

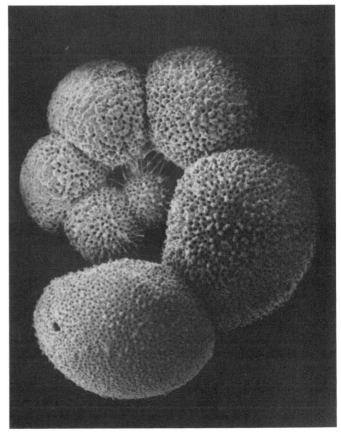

Fig. 3 × 125

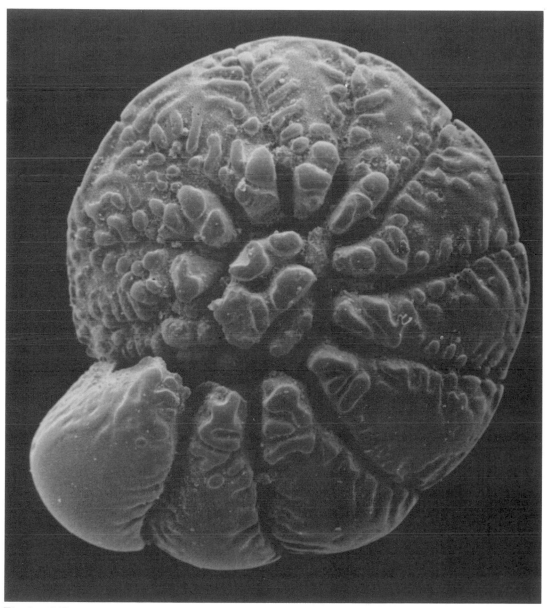

Fig. 4 × 140

HYDRA

Fig. 1 × 41

Hydra is a simple multicellular organism, about 1/4 inch long. It is commonly found in pond water attached to hard surfaces such as plants and rocks. The main part of its body is a hollow flexible tube, which is closed at one end. The open end, which is its mouth, is surrounded by five tentacles (Fig. 1). The tube serves as the digestive cavity and is lined with cells which secrete digestive enzymes. The closed end of the tube forms a flattened plate and is used to attach the animal to various surfaces. Hydras are usually stationary, spending most of their lives in one or a small number of places. However, a hydra is capable of moving. It accomplishes this in an unusual way—by bending over and attaching to a second location with its tentacles (Fig. 2), releasing its foot and then reattaching the foot at a new place. This means of locomotion is called *somersaulting*. Hydra has primitive nerves and muscle-like cells and can use these cells both for somersaulting and for changing its shape (compare Figs. 1 and 3). Near the base of the tentacles is the mouth of the organisms. The mouth is barely visible in Fig. 4 but is shown magnified in Fig. 5. Embedded in the tips of the tentacles are specialized capsules containing stinging units called *nematocysts*. Two of these can be seen as small spheres in Fig. 6. When appropriate prey contact the capsules, trigger hairs cause the nematocyst to discharge.

Fig. 2 × 125

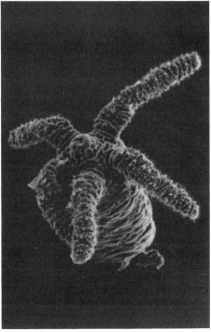

Fig. 3 × 125

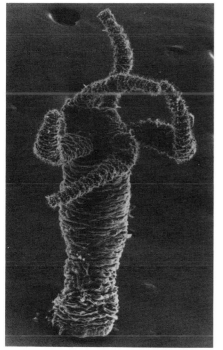

Fig. 4 × 105

A nematocyst is shown at two different magnifications in Figs. 7 and 8. A long poisonous thread can be seen. The thread penetrates the prey and releases a poison. Small animals such as protozoa or tiny insects are paralyzed by this poison. Following discharge of the nematocyst, the tentacles bring the paralyzed prey to the mouth. Hydra can even consume organisms larger than itself; a part of the organism is put into the mouth and as digestion proceeds, more of the organism is pushed in. Hydras have a remarkable ability to repair themselves after receiving a considerable amount of mechanical damage. For example, if a hydra is chopped up into many small fragments, each of the fragments begins to enlarge. This continues until each fragment has regenerated a complete organism. Hydras also reproduce in a rather unusual way. Occasionally a small bump appears on the side of a mature hydra. The bump enlarges, forming a cylindrical appendage. Tentacles form and in time the newborn hydra breaks off from the body of its parent. It quickly finds a suitable site of attachment and begins to grow. Often one parent hydra can have several baby hydras growing at the same time. Hydras also have a sexual mode of reproduction but the branching method is by far the more common.

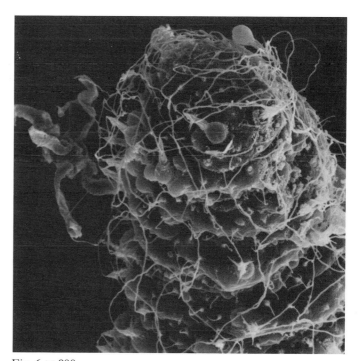

Fig. 6 × 900

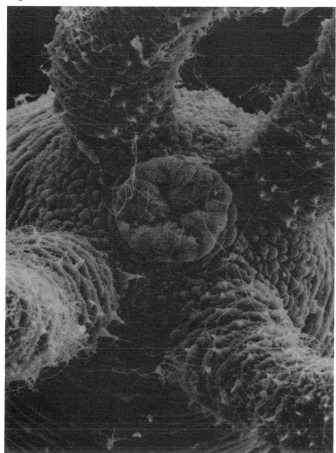

Fig. 5 × 320

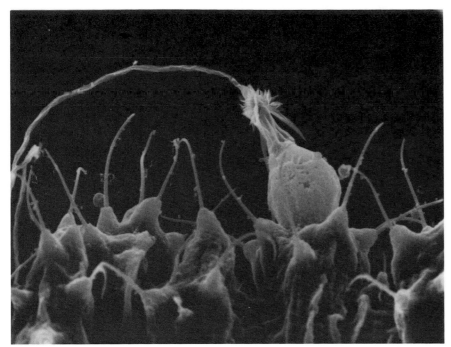

Fig. 7 × 3,300

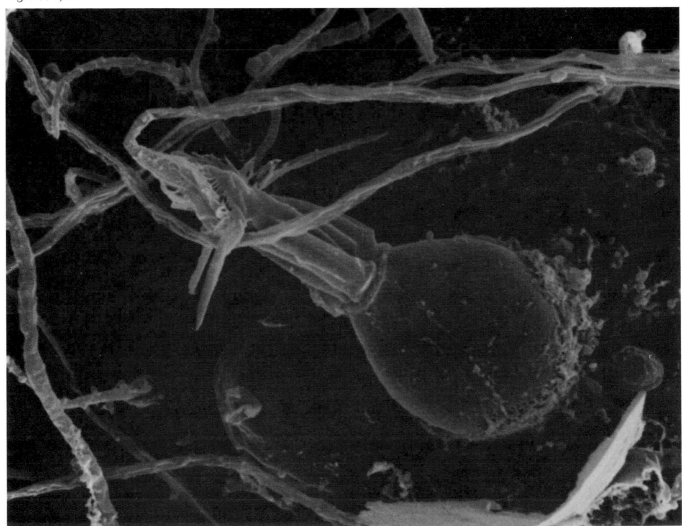

Fig. 8 × 8,120

SPINY-HEADED WORMS

Fig. 1 × 72

The spiny-headed worms are tiny parasites (about 1/30 inch long) found in the digestive tract of a variety of vertebrates. The animal shown in the figures was obtained from the wall of the small intestine of a rat. Fig. 1 shows the complete animal. The front part of the worm (its "head"), which is magnified in Fig. 2, contains several spines, arranged symmetrically. Some of the spines are curved in such a way that the animal can hook itself to the inner wall of the small intestine of the host. Here it spends most of its life feeding on nutrients passing through the intestine.

Fig. 2 × 835

TAPEWORM

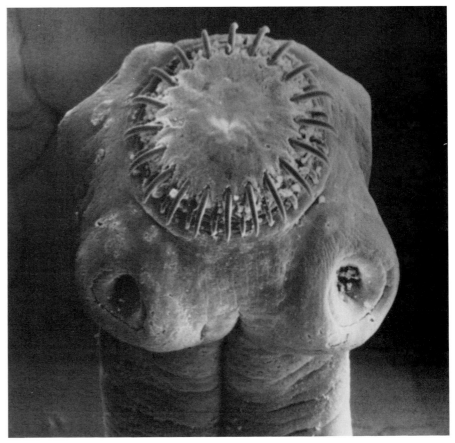

Fig. 1 × 90

Tapeworms (Fig. 1) are parasitic flatworms found in the intestine of such warm-blooded vertebrates as birds and mammals. The front portion (the "head") of the worm (Fig. 2) contains four rounded, cup-shaped suckers and a crown of hooks, which are used to attach the worm to the intestinal wall. The animal feeds on the contents of the intestine and can grow to several feet in length. Their skin is resistant to the digestive enzymes of the host. Tapeworms have an unusual life cycle. Eggs develop in the hindmost segment, which breaks off and is expelled in the feces of the host. The eggs are ingested by other animals in whose intestines they hatch. The young worms bore through the intestinal wall to blood vessels and are carried to muscles in which they become dormant. If uncooked meat of this host is eaten, for example, by a human, a worm may be released in the human intestine, where it attaches to the intestinal wall and starts a new life cycle.

Fig. 2 × 110

BLOOD FLUKE

Fig. 1 × 38

Fig. 2 × 925

Blood flukes are flatworms responsible for schistosomiasis, a disease found in some tropical regions. The worms spend part of their lives in the blood vessels of mammals—hence the name. The blood fluke lays eggs in a tiny blood vessel in the intestinal wall. The vessel ultimately ruptures, releasing the eggs, which are discharged in the animal waste. In modern sewage systems the eggs are destroyed but in countries where raw sewage is used as fertilizer, the eggs hatch and larvae bore into the body of certain snails, where they multiply. The new worms are released from the snail and if one contacts another animal (for example, a human walking barefoot through a rice paddy), it bores its way through the skin and starts a new life cycle. The female blood fluke is generally not motile but remains in a groove on the male (Fig. 1). The male worm has many rounded projections on its body (Fig. 2). These projections are covered with spines with which the worms hold on to the walls of the intestinal blood vessels.

SNAIL

Fig. 1 × 3

Snails are animals that spend their lives within a coiled shell (Fig. 1). The shells range in size from 1/16 inch to 3 inches. The animal has a flattened muscular foot and a definite head with a mouth and two tentacles which bear its eyes. The mouth, which usually maintains contact with the surface on which the snail crawls, contains a special rasping organ called the *radula*. Numerous "teeth" are located on the tongue-like structure, arranged in long parallel rows. Several of these rows are illustrated in Fig. 2. When a snail feeds, it holds food in the fleshy jaws surrounding the radula and rasps away with these teeth. The radula moves forward and back in a wave-like motion. If the food particles are attached to a surface, such as algae or the surface of a leaf, the radula is pushed out of the mouth and it scrapes the food from the surface. When so doing, the snail almost always glides across the surface as it feeds.

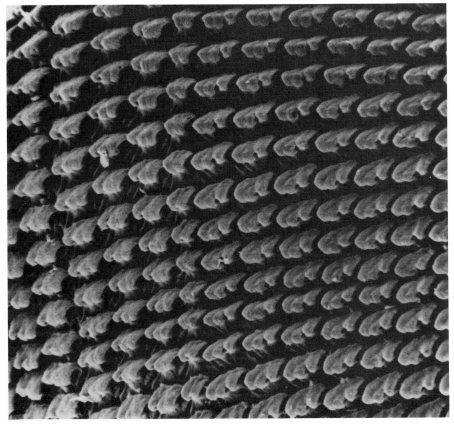

Fig. 2 × 780

Fig. 1 × 64

INSECT EYE

The head of a fly is a multicomponent structure containing a complex mouth and several sensory organs (Fig. 1). The two large compound eyes seen in the figure, one on each side of the head, are the most striking feature of the head. Each eye is composed of smaller units arranged in an orderly geometric pattern, which can be seen in the enlargement in Fig. 2. Higher magnification (Fig. 3) shows that these units are closely packed together. Each unit has a separate lens system; this arrangement gives the insect a wide field of vision, which is valuable in avoiding enemies. The hairs observed between the units are vibration-sensitive elements and may enable the fly to detect nearby insects and other animals.

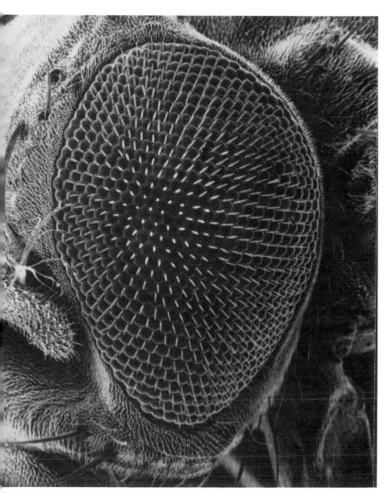

Fig. 2 × 230

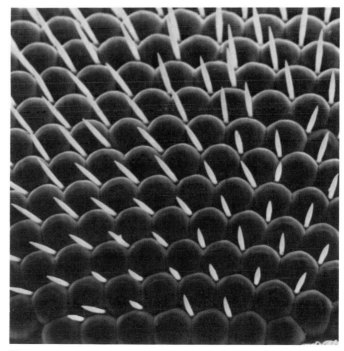

Fig. 3 × 740

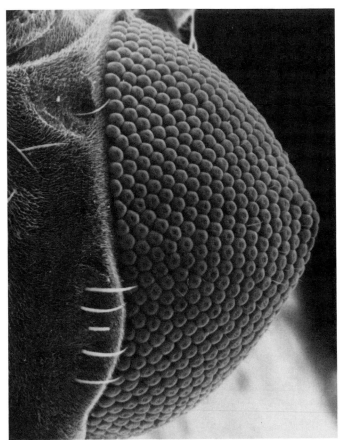

Fig. 4 × 280

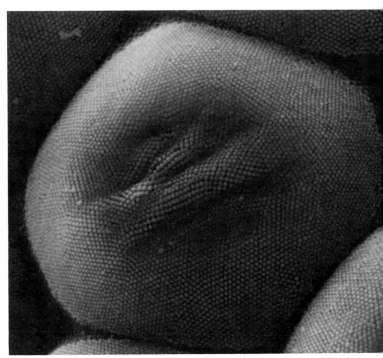

Fig. 5 × 6,425

The eyes of the moth are also large with many facets (Fig. 4). Each facet has a hexagonal shape, as can be seen in the low-magnification photograph in Fig. 4. The enlargement of a single facet (Fig. 5) shows that each facet is covered by many small projections. A common question about the insect eye is how an insect having thousands of lenses in its compound eyes manages to see a single image. This situation is of course no different from the human, which forms a single image from two eyes. The insect brain, like the brains of two-eyed animals, processes the information provided by the eyes to form a single sensory image. Interestingly, the light-sensitive pigment of insect eyes is nearly identical chemically to vitamin A, which is required for vision in mammals.

INSECT MOUTH
Housefly

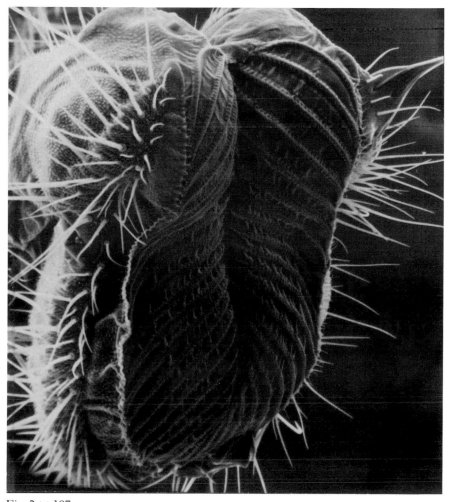

Fig. 1 × 35

A front view of a fly's head is illustrated in Fig. 1. The two, large eyes occupy much of each side of the head. Extending from the very front is an appendage called the *proboscis* (Fig. 1). The tip of the proboscis is magnified in Fig. 2. The proboscis is a flexible, hollow tube, the end of which has two rather large lobes. These lobes contain grooves or canals which connect with the central tube. When the fly feeds, it extends the proboscis to the food and if the food is a fluid, it enters the openings of the lobes by capillary action. Many long, needle-like hairs surround the tip of the proboscis. These are sensory organs used to "taste" various materials with which the proboscis comes into contact.

Fig. 2 × 197

INSECT MOUTH

Butterfly

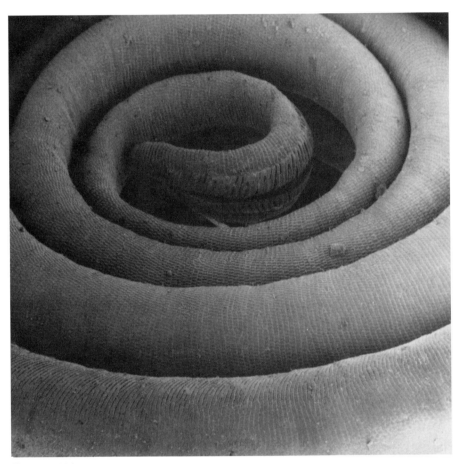

Fig. 1 × 16

Fig. 2 × 130

A portion of the head of a butterfly is shown in Fig. 1. The large oval and smooth structure seen in the figure is one of the two compound eyes. The coiled structure is the *proboscis,* or sucking organ. It is shown in its resting state. When the butterfly feeds, it uncoils its long proboscis and places it in a droplet of liquid food. The fluid is then sucked into the oral cavity of the head. In the enlargement of the proboscis (Fig. 2), it can be seen that the tip consists of two parts, each having a groove, and that the parts are positioned together in such a way as to form a tube. These components, called *maxillae,* can be separated in order to clean the proboscis and remove particles that may have clogged the tube. In other insects the maxillae are more motile and are used to bring solid food into the mouth.

INSECT ANTENNAE
Moth

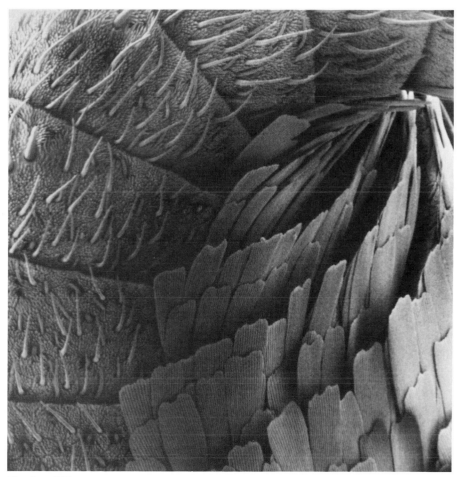

Fig. 1 × 84

Many insects have sensory structures called *antennae*, commonly called "feelers." The antenna of a moth is shown in Fig. 1. On many insects the antennae are like fine feathers but on the moth the antenna has flattened, overlapping scales and many tapering, rod-like structures which are the sensory receptors. The magnified view shown in Fig. 2 illustrates that these receptors differ in shape, number, and distribution. These differences correspond to differences in function. Some of the sensory receptors detect movement and materials, whereas others are used to detect odors. The antenna can be moved by muscles at the base. Vision in moths is poor and the moth probably detects little other than light and dark. Consequently, antennae provide the more reliable information about the surroundings.

Fig. 2 × 760

INSECT ANTENNAE
Mosquito

Poised to feed, a mosquito stands ready to stab its proboscis through the skin of an animal. Should it escape a responding slap, it will obtain a droplet of blood that can provide nourishment to the mosquito for up to three days. The elaborate object shown in Fig. 1 is part of the "face" of a mosquito. Two long antennae situated between the eyes take up much of the surface of the head. The lower portion of one of these antennae is shown in the figure. At its base is a large doughnut-shaped structure which is a muscular organ used to move the antenna. Adjacent to the antenna are clusters of hairs which serve a sensory function. A high magnification view of the central part of the shaft of a single antenna is shown in Fig. 2. The many spines are sensory elements responsible for smell and touch. A portion of the mouth is also shown in Fig. 1; the blood-sucking tube emerges from the center of the mouth and is pointed to the lower right.

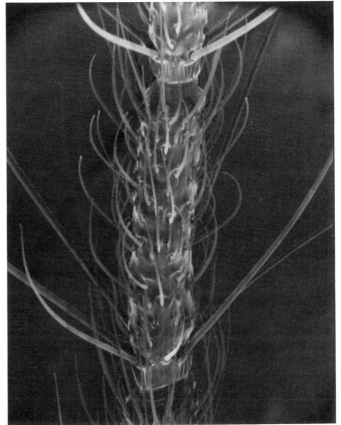

Fig. 1 × 133

Fig. 2 × 515

92

INSECT WINGS

Butterfly

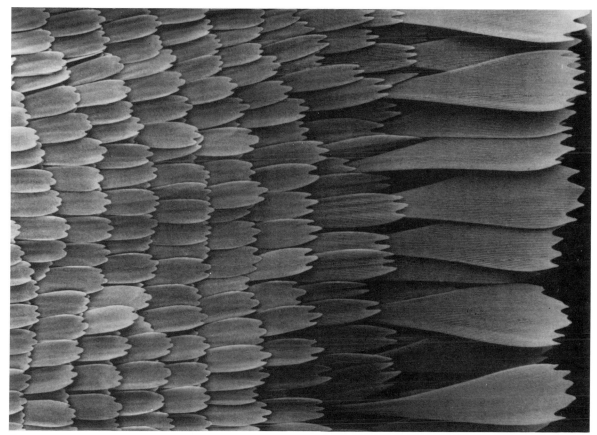

Fig. 1 × 2/3

A butterfly stands lightly on a flower in Fig. 1, holding its wings erect. (This posture is one way to distinguish moths and butterflies, for moths lay their wings flat when resting.) Butterfly wings are covered with scales having a variety of shapes (Fig. 2). These scales can easily be brushed off when one holds a butterfly wing and are commonly called "dust." In Fig. 3 two scales are magnified to show how each scale is inserted into a socket in the wing membrane; other scales are under the two that are shown. The scales overlap so that they form a complete covering for the wing. Pigmentation of the scales accounts for the glorious colors of butterflies. A portion of one scale is magnified and shown in Fig. 4. Note that the scale is also a complex structure, consisting of small overlapping plates that form ridges and are linked to adjacent scales by cross-bridges. This arrangement strengthens the scale without adding to its weight.

Fig. 2 × 140

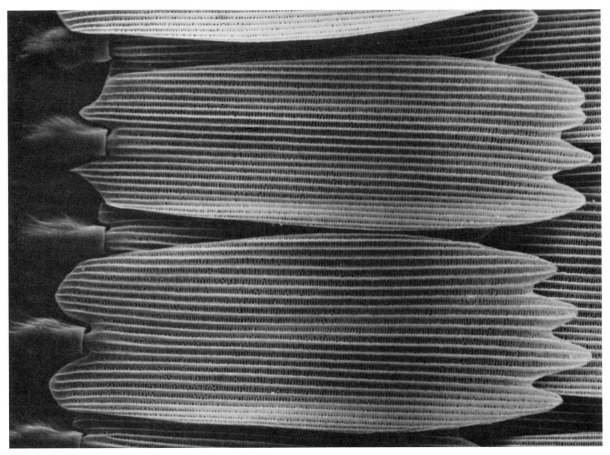

Fig. 3 × 1,220

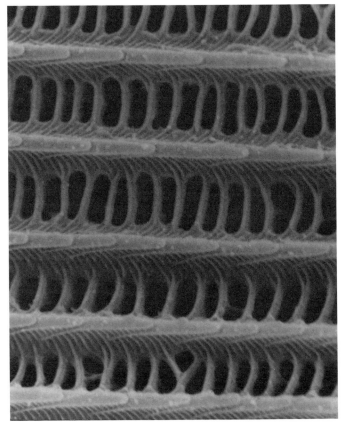

Fig. 4 × 12,400

INSECT WINGS

Lacewings

Fig. 1 × 34

Lacewings are delicate insects having four equally-sized wings that are often green. Each wing has many veins that make interesting patterns (Fig. 1). A part of a single vein is also shown at high magnification in Fig. 2. Notice how the veins are connected to many needle-like hairs on the wing surface. These hairs prevent the delicate surface of the wings from being damaged by small particles. Lacewings are among our most valuable insects for they feed voraciously on aphids and various scale insects that damage fruit trees. Releasing lacewings in an orchard can be used as a substitute for chemical spraying.

Fig. 2 × 290

INSECT EXOSKELETON

Fig. 1 × 55

The hard-bodied insects (for example, roaches, flies, and mites) are encased in a suit of armor called an exoskeleton. In the early development of the insect the exoskeleton forms by secretion of a liquid which quickly hardens to form a smooth "shell." Since the exoskeleton cannot grow, the insect must shed it periodically as the insect grows. Sometimes the exoskeleton is divided into plates to provide flexibility. The exoskeleton serves two functions. First, it is a protective skin which prevents damage and dehydration. Second, insect muscles are attached to the exoskeleton just as muscles are attached to bones in the higher animals. The exoskeleton is usually elaborately decorated. Fig. 1 shows the head of a fruit fly; the spiny exoskeleton just behind the eyes is magnified in Fig. 2. Occasionally long slender hairs are set in the exoskeleton surface. These hairs can be seen in the exoskeleton of the housefly (Fig. 3) and the red mite (Fig. 4). The surface of the collembola (Fig. 5) is hairless and quite lovely.

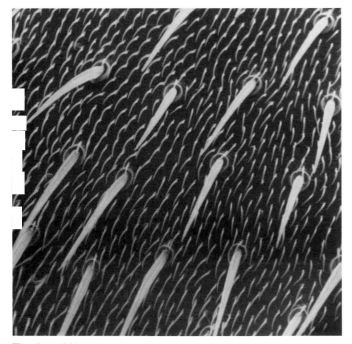

Fig. 2 × 560

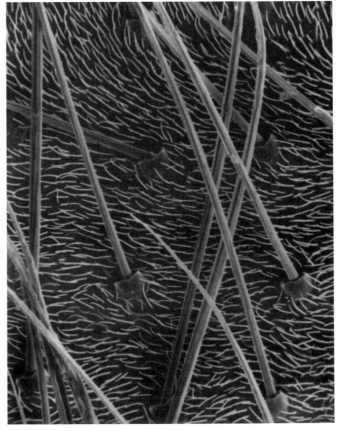

Fig. 3 × 695

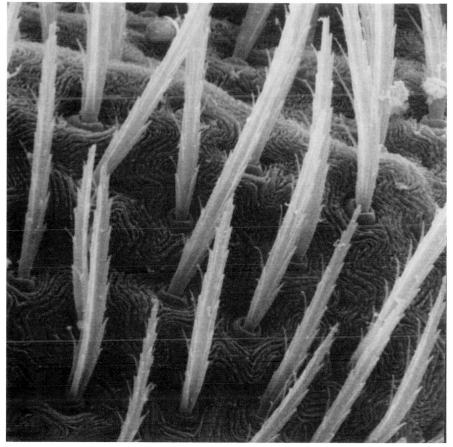

Fig. 4 × 2,170

Fig. 5 × 9,440

HONEYBEE

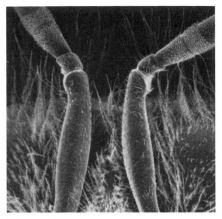

Fig. 1 × 35

The common honeybee is a gentle animal when handled carefully. It is useful for man and animals as a source of delectable honey which it manufactures from flower nectar. However, it is far more important in nature as a major transporter of pollen from one flower to another. Many plants, especially fruit trees, are unable to produce seeds and fruit without this pollination. The head of the honeybee (Fig. 1) has two large eyes and adjacent to the eyes are two long segmented antennae. A portion of the two antennae, just after they emerge from the head, are shown. Hairs cover the entire head. A closeup of an antenna is shown in Fig. 2; the boundary between adjacent segments is shown, as well as many plates and spines, some of which are for touch and some for smell. The inner surface of the hind leg of a worker honeybee is shown in Fig. 3. It is covered with numerous spines that are used to collect pollen. Some pollen grains can be seen in the figure. By a complex movement of the legs and the two spines, a worker bee is able to form a ball of all of the pollen grains and deposit it in a structure on the hind legs called a *pollen basket*.

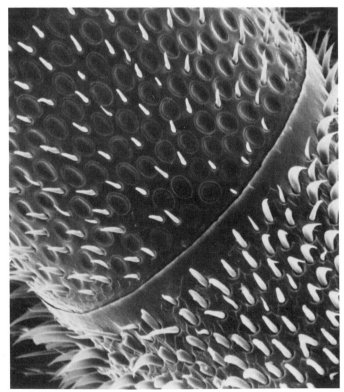

Fig. 2 × 500

Fig. 3 × 430

Fig. 1 × 8

BEETLES

Ladybug

The body of a beetle is not only encased in an exoskeleton but its upper surface has a divided hardened cover (Fig. 1) that protects the delicate hindwings. The covers are elaborately colored in some beetles. The underside of the ladybug (Fig. 2) has three pairs of legs and is covered with fine hairs. Two short antennae emerge from the front part of the body (Fig. 2). The legs are terminated with claw-like structures, which enable them to hold on to a variety of surfaces.

Fig. 2 × 32

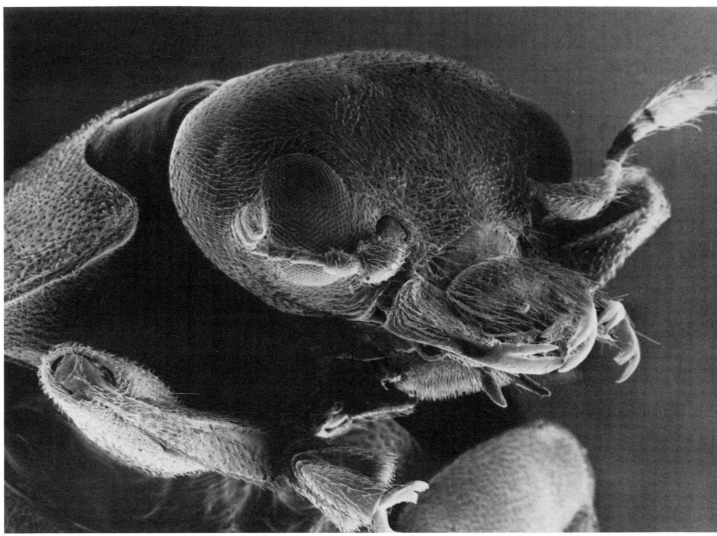

Fig. 3 × 65

On this page is an enlargement of the head of a ladybug (Fig. 3). The eyes are smaller with respect to the head than those of the flies seen earlier. This is probably because of a reduced need to see predatory enemies; the ladybug secretes a liquid with a tobacco-like odor that is repulsive to other insects and birds. Note the multicomponent mouth which is useful for chewing and crushing other insects. A clawed foot can be seen clearly at the bottom center of the electron micrograph. Ladybugs are among the most useful insects because they feed voraciously on common plant pests, such as mites, aphids, and scale insects. A single ladybug larva, sometimes called an *aphid lion* or *aphid tiger*, can consume two hundred aphids in a single day.

MITES

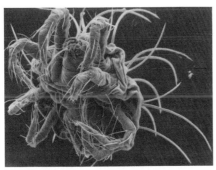

Fig. 1 × 155

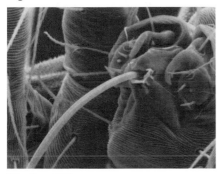

Fig. 2 × 715

Mites are in the spider family and thus have four pairs of legs (Fig. 1) in contrast with insects, which always have three pairs. Most mites are also covered with a large number of hairs, as shown in the figure. A part of the mouth is shown in Fig. 2. Notice the long spine projecting leftward from the mouth. This spine is used to pierce the underside of leaves. Mites have sucking mouths and feed by sucking juices from the plant. In so doing they can do considerable damage to plants. An infestation of house plants is especially damaging as there are no natural enemies in the home and the mites multiply rapidly. Fig. 3 shows the elaborately folded and sculptured exoskeleton from which numerous sensory hairs project. Many species of mites are covered with a fine web which protects them from insect sprays unless the spray is very forceful. The best protection against mite damage in the garden is to infect the garden with lady bugs, which feed on mites. In the home a soap solution may be used to kill mites.

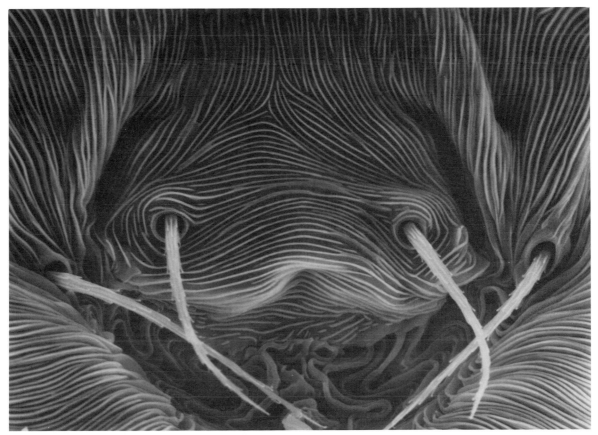

Fig. 3 × 2,880

MILLIPEDES

Fig. 1 × 1

Millipedes are multi-jointed animals (Fig. 1) having a pair of legs on each segment of the body. They are often called "thousand leggers." The head has a delicate pair of short antennae (Fig. 2). The mouth, which is located on the underside of the head, has many parts, as is common with chewing insect-like animals. Millipedes, though frightening looking, are harmless animals, feeding at night on bits of wood and various organic materials found in soil. In contrast, centipedes, which have a similar appearance, have a poison gland and feed almost exclusively on insects.

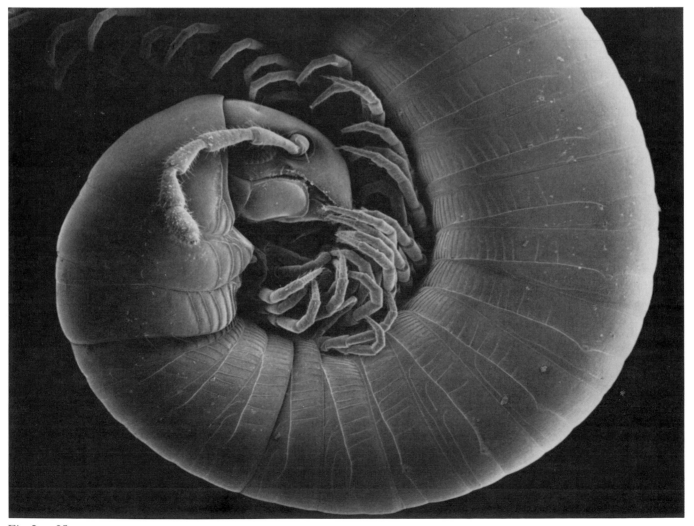

Fig. 2 × 25

SNAKE SKIN

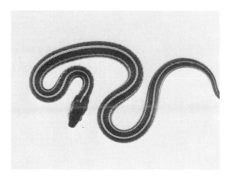

Fig. 1 × 1/4

Snakes (Fig. 1) are a group of animals without limbs. Locomotion in snakes is achieved by wavelike contractions of its body. To prevent slipping backward, snakes are covered with overlapping horny scales that project backward (Fig. 2, to the right). These scales are bone-hard but are not bone; rather, they are skin cells which have fused and developed a hard layer. Because snakes are very long, they have a large surface area for their weight and thus are very susceptible to drying. The scales are impermeable to water and thus reduce the rate of drying; still, snakes are unable to remain in bright sunlight for long.

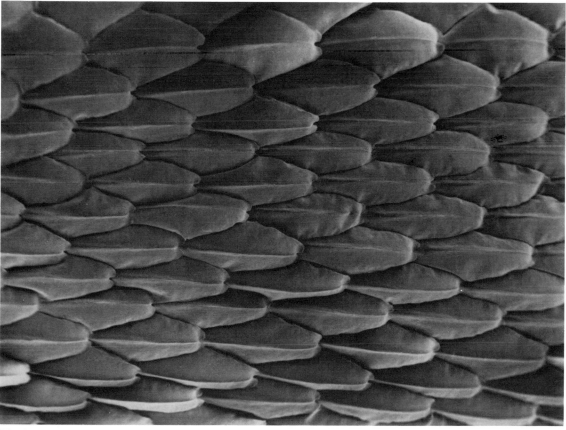

Fig. 2 × 30

SNAKE TONGUE

Because snakes lack limbs, they have developed an agile tongue to bring small particles of food into the mouth. For the common garter snake, whose tongue is shown in Fig. 1, the tongue is long, tapered, and forked. The tongue of the snake has many functions other than that of acquiring food. For example, its sensitive surface is used as an accessory organ of smell. The constant ejection and withdrawal of the tongue is the means by which the snake smells the air around it. The broad surface of the tongue is covered with elongate, flattened cells, which can be seen in Fig. 2. The base of the tongue is more like that of the tongue of higher animals and has a folded surface covered with numerous projections (Fig. 3). Many areas of the base of the tongue are covered with sensory cells possessing hair-like appendages called *cilia* (Fig. 3), much like the cilia seen on the surface of some protozoa (pages 70–74). The cilia are probably able to sense vibration. In this respect, the tongue may compensate in part for the lack of ears.

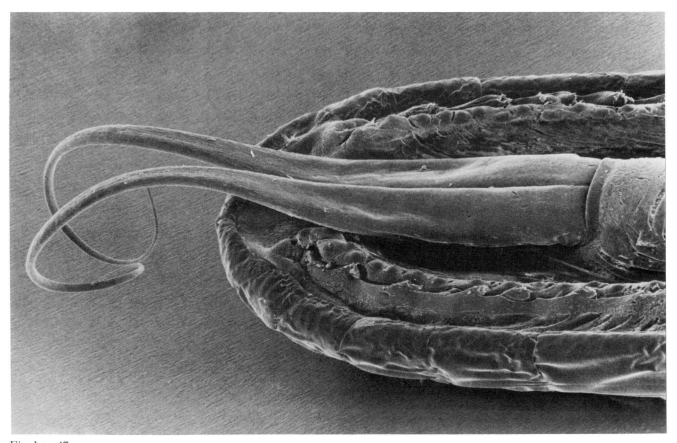

Fig. 1 × 47

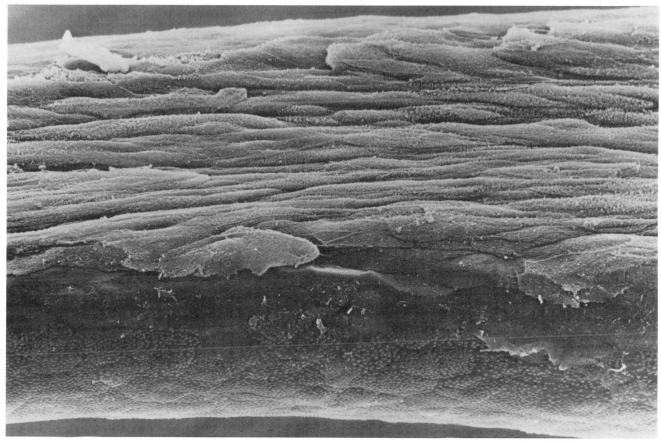

Fig. 2 × 800

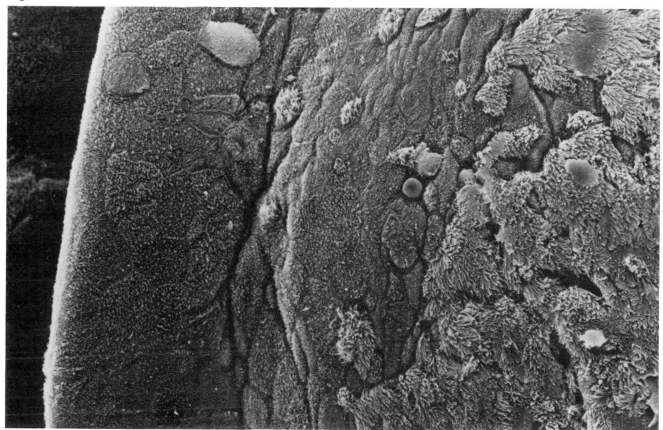

Fig. 3 × 765

FEATHERS

Sparrow

Fig. 1 × 1

Fig. 2 × 75

Feathers (Fig. 1) form a light covering that protects birds from water and from heat loss. Feathers, like snake skin (see previous page), are derived from skin cells. In contrast with snake skin scales, feathers are also able to be moved by muscles under the skin. Some birds have several thousand feathers, each of which grows from a localized region of the skin. Each feather consists of a long, central shaft (Figs. 1, 2) that becomes a hollow quill just before the shaft enters the skin. Many barbs extend from the main feather shaft and small branches (barbules), in turn, extend from each barb (Fig. 3). The ends of the barbules terminate as small hooks, shown in Fig. 4. A thin transparent vane (Figs. 2, 3) is associated with each barbule. A system of hooks at the ends of the barbules keeps the feathers joined together. When the barbules are hooked, the vanes of adjacent barbules overlap, forming a surface through which air cannot easily pass during flight and also preventing strong winds from disrupting the feather structure.

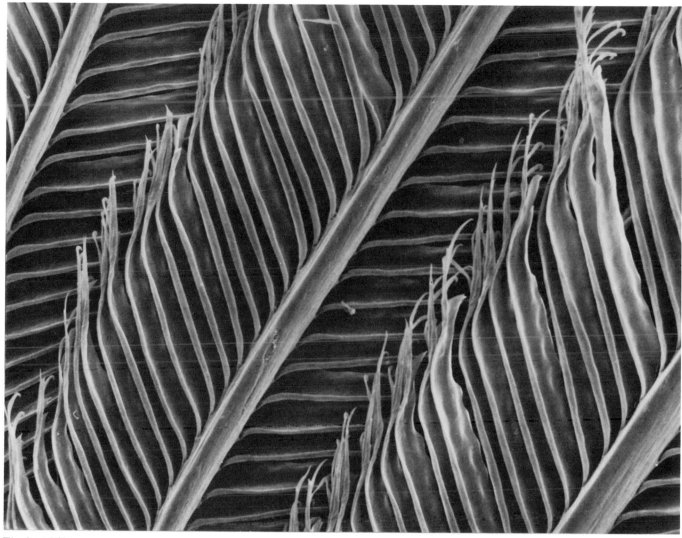

Fig. 3 × 360

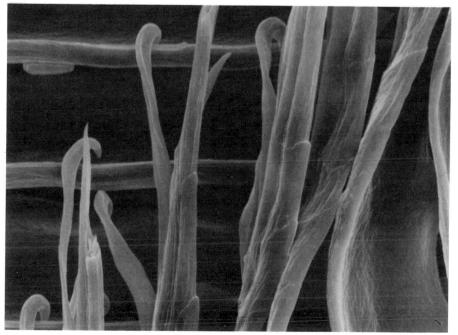

Fig. 4 × 1,800

CELLS IN DIVISION

All animal and plant cells multiply by cell division. This process, which is called *mitosis*, occurs in a series of distinct steps. First, the membrane enclosing the nucleus of a cell that is ready to divide disintegrates and the genetic material (DNA) is released to the remainder of the cell. Shortly afterwards, the DNA condenses with special proteins to form *chromosomes*. All of the chromosomes then line up near the middle of the cell. There are two sets of chromosomes at this time and, by a complex process, each set is guided to opposite sides of the cell. Following separation of the chromosomes, the cell then divides. The beginning of mitosis can be recognized either by the dissolution of the nuclear membrane or by an associated rounding-up of the cell. The shape changes and several stages in the splitting of a cell are shown in these panels. Fig. 1 shows a plastic surface on which many growing cells are attached and spread out. The cells have a variety of very noticeable projections, whose functions are not known. In this figure two nearby cells are in the early stage of cell division and have rounded up. Another cell in an early stage is shown in Fig. 2 at higher magnification. The cell in Fig. 3 is elongated; presumably the two sets of chromosomes (not visible because we are looking only at the surface of the cell) have already separated. The cell in Fig. 4 is at a more advanced stage of division than the one in Fig. 3. The cell in Fig. 5 has nearly divided. Once division is completed, the two daughter cells again become spherical (Fig. 6). Shortly after completion of division, the chromosomes expand and become more diffuse until they are no longer evident. A nuclear membrane reforms in each cell and the cells then flatten out again on the plastic surface and resume growth.

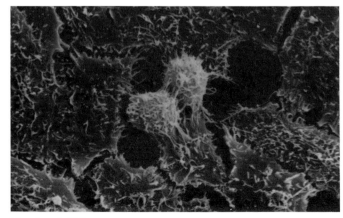

Fig. 1 × 1,200

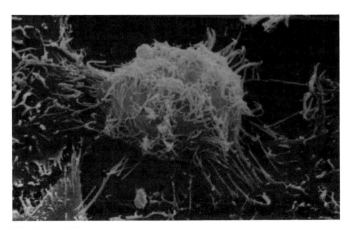

Fig. 2 × 2,250

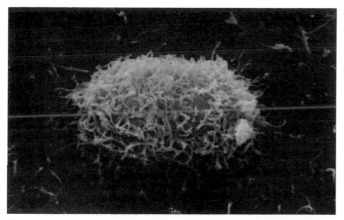

Fig. 3 × 2,400

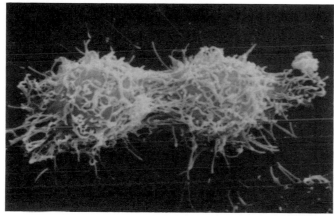

Fig. 4 × 2,300

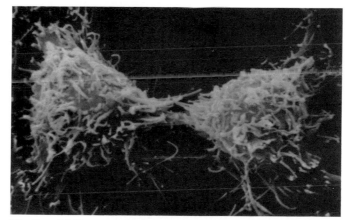

Fig. 5 × 2,360

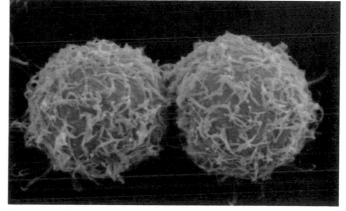

Fig. 6 × 2,410

BLOOD

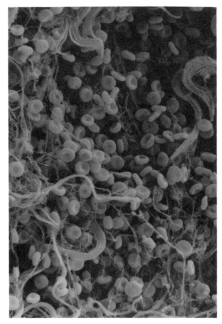

Fig. 1 × 1,050

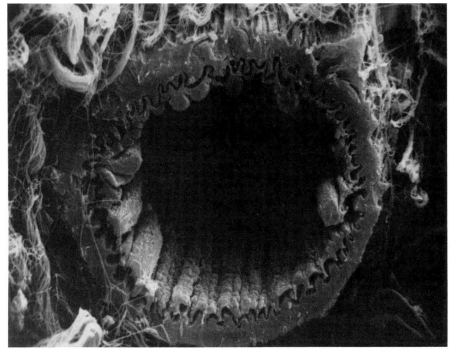

Fig. 2 × 2,475

Blood is an enormously complicated substance. It is primarily a concentrated solution of various salts and proteins in which there is suspended a variety of cells. Furthermore, blood carries oxygen and nutrients to all body tissues and removes carbon dioxide and all other waste products, carrying them to the lungs or kidneys. In addition, elements in the blood are responsible for recognizing foreign matter, such as bacteria and viruses, synthesizing antibodies, and for destroying and removing the foreign material. The most noticeable cell in blood is the red cell or *erythrocyte*. Many of these are shown in Fig. 1. These are discs which are hollow on both sides, as can be seen in Fig. 4, which includes one red cell at a higher magnification. The color is a result of *hemoglobin*, the protein that carries oxygen. Blood circulates through the body through a branched closed network of blood vessels. A cross-section of one blood vessel is shown in Fig. 2. This is a thick-walled artery, consisting of a muscular outer wall and an inner lining of cells in folds. If a blood vessel breaks and blood is exposed to air, a protein called *fibrin* is released. This protein aggregates forming a fibrous clot in which many red cells are trapped. Fig. 1 shows some clotted fibers.

Blood contains many white cells, five of which are shown passing through a vein in Fig. 3. These cells, which are considerably larger than red cells, are primarily responsible for protecting the animal against foreign organisms by synthesis of antibodies. Some of the cells recognize the foreign matter, others signal the immune system to act, and others synthesize antibodies. The foreign material is usually precipitated by the antibodies, after which the largest white cells ingest the precipitate and destroy it. Blood *platelets*, the irregularly shaped fragments in Fig. 6, are responsible for clotting.

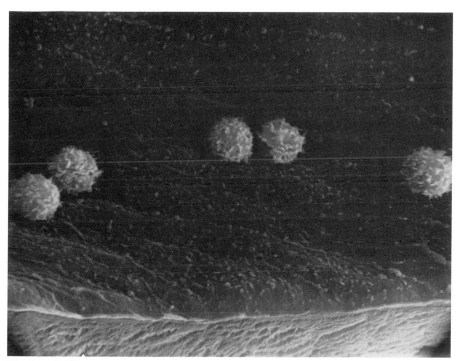

Fig. 3 × 2,685

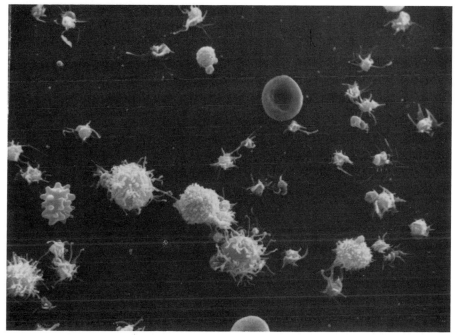

Fig. 4 × 2,655

BONE

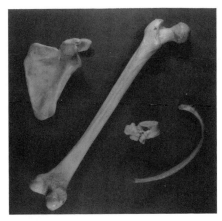

Fig. 1 × 1/7

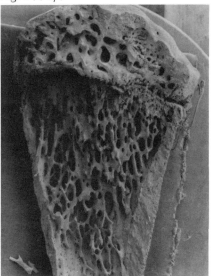

Fig. 2 × 9

Bone is a specialized tissue designed to bear weight. It comes in a variety of shapes (Fig. 1). Bone consists both of cells and of specific products that the cells have synthesized and released to their surroundings. The first products made are tough protein fibers and complex sugar-like polymers. Soon after these substances are released from the bone-forming cells, the structures formed by these molecules are impregnated with calcium salts; this calcification process gives bone its hardness. The end of the bone of the leg of a rat is shown in Fig. 2. At low magnification (Fig. 1) the bone appears solid but at the higher magnification of Fig. 2, one can see that the bone is very porous. This porosity is characteristic of the ends of most long bones. The shaft of bones is usually much more compact. Fig. 3 shows an enlargement of a portion of the shaft. The larger round holes are passages for blood vessels. The small holes are spaces in which a single cell formerly responsible for synthesizing the bony substances resides. When actively making bone, the cells are called *osteoblasts* (*osteo*, bone; *blast*, maker). Once the cell is entrapped by its secretory product, that is, the bone substance, these cells stop producing bone and are called *osteocytes* (*osteo*, bone; *cyte*, cell). Fig. 4 shows an enlargement of one of these spaces. The sponge-like object is the osteocyte. The area surrounding

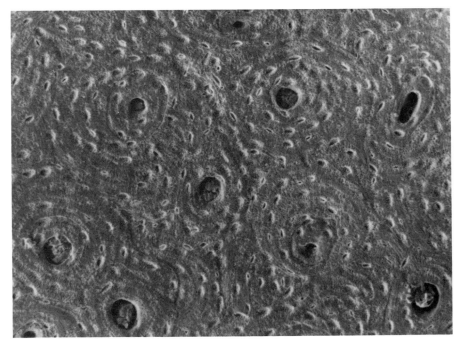

Fig. 3 × 185

the osteocyte can be seen, on close examination, to consist of numerous fibers and small crystals of calcium salts. (The composition of these crystals is quite similar to that of the enamel of teeth.) The shaft of many of the long bones contains a central core which contains a pasty material called bone marrow. This material is a mixture of cells which differentiate to form the various components of blood and large cells responsible for forming fat. The fat cells usually reside in pockets in the wall of the marrow cavity. Some of these pockets are shown in Fig. 5. Each pocket indicates the position of a single fat cell.

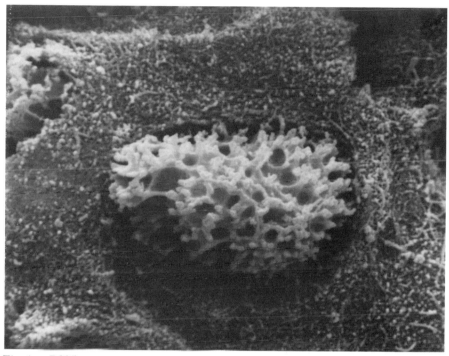

Fig. 4 × 7,215

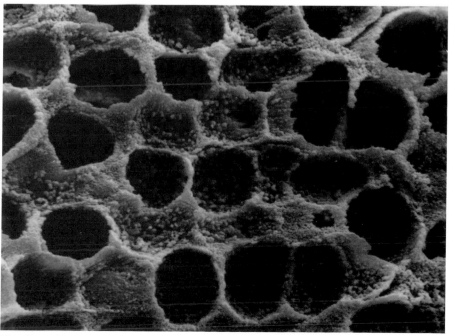

Fig. 5 × 855

LYMPH NODES

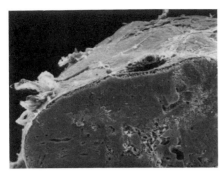

Fig. 1 × 57

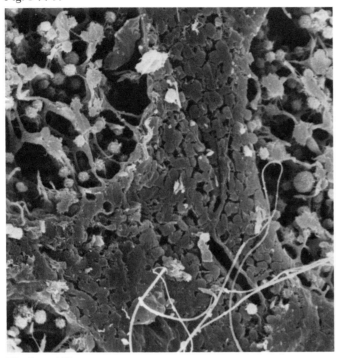

Fig. 2 × 668

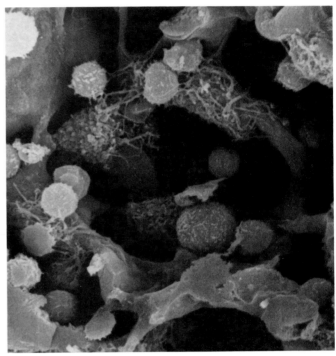

Fig. 3 × 2,270

In addition to blood, a clear colorless liquid called *lymph* flows through the body in lymph vessels. At numerous points in the body the lymph vessels enter a dense network known as a lymph node. The lymph node is charged with filtration of body fluids and removing foreign material from the body. Dissection of a lymph node (Fig. 1) shows that it consists of very closely packed clusters of cells surrounded by small tubes and a fairly loose network of cells. Details of the loose network, which is called a *sinus*, are shown in Figs. 2 and 3. Several different cell types can be distinguished in Fig. 2 by the size and shape of the cells. Most of the cells are closely related to the white blood cells and have a similar function, namely, to synthesize antibodies. Many foreign proteins and especially precipitates and aggregates are trapped in the filter-like lymph nodes. Antibodies are made, which cause further aggregation. A common cell in lymph nodes is the macrophage ("big eater"). These huge cells, three of which are in Fig. 3, are able to ingest particulate matter such as the protein aggregates. The macrophages are recognizable in the figure both by their size compared to the small round cells (which produce antibodies) and by the long slender extensions coming from their surfaces.

SKIN

The skin and other structures such as hair, sweat glands, oil glands, and nails enclose the body. The principal function of the skin is protection, both against mechanical damage and against drying. A secondary function is sensory, that is, as a system containing the nerve endings responsible for the sense of touch and temperature. In animals the hair serves also for protection and to keep the animal warm. A section through the skin of the ear is shown in Fig. 1. Numerous hairs extend through the outermost layer of the skin in the right part of the figure. To the right of center are bundles of nerve fibers.

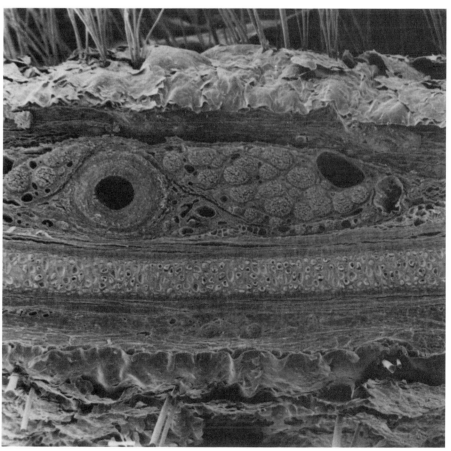

Fig. 1 × 150

The bundles of fibers seen in Fig. 1 are enlarged in Fig. 2. Each circular unit in the bundle is the cross-section of a single nerve cell surrounded by a fatty sheath. The sheath is an important electrical insulator between adjacent nerve cells. Running vertically at the left of Fig. 1 is a cartilage band. This band is magnified in Fig. 3. Like bone, cartilage is formed from the secretions of cartilage-producing cells called *chondroblasts* (*chondro*, cartilage; *blast*, maker). The individual units are the clusters of chondroblasts. They reside in small cavities as in bone; the space between the clusters is the cartilage.

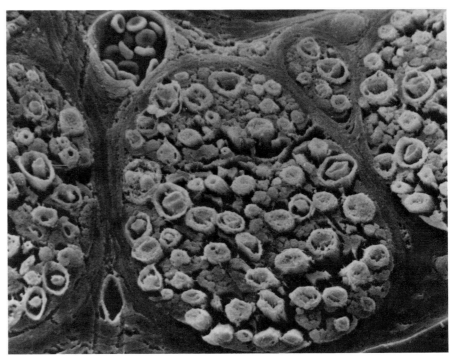

Fig. 2 × 1,200

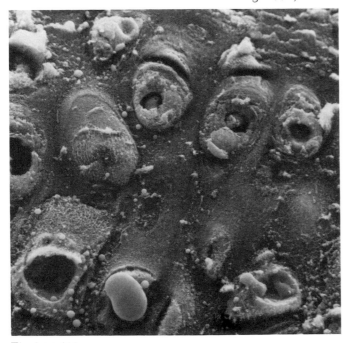

Fig. 3 × 544

SKIN

Fig. 1 × 8

The outermost layer of the skin, the *epidermis*, has particular patterns of folds and ridges. These are most noticeable in the fingerprints (Fig. 1). A higher magnification of the center of the fingerprint shows that the ridges themselves are highly wrinkled. The centermost portion of Fig. 2 is magnified even further and shown in Fig. 3, which indicates that even the wrinkled folds are wrinkled. This area is enlarged in Fig. 4, which shows the opening of a sweat gland. The cells of the epidermis are continually being abraded away. Cells in the lower layers (the *dermis*) reproduce and push to the surface to replace the epidermal cells. The outermost cells are not alive. The area surrounding the sweat pore in Fig. 4 includes several loose cell fragments which are getting ready to slough off. In some cases the skin contains protein fibers which hold dead cells together so that the skin sloughs off in flakes; this is common on the scalp.

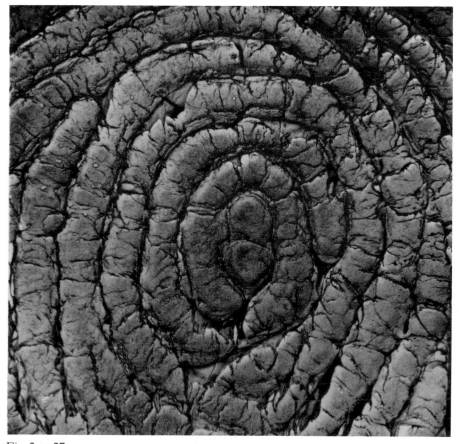

Fig. 2 × 27

117

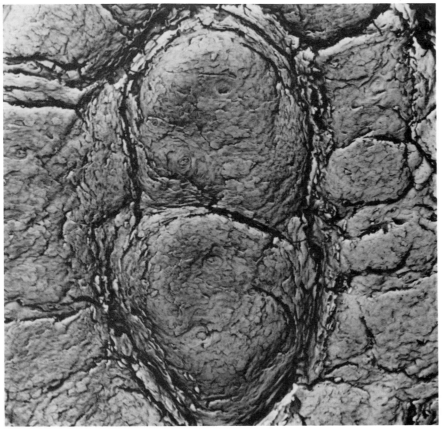

Fig. 3 × 106

Fig. 4 × 1,795

118

HAIR

Hair has two parts—the root and the visible exterior shaft. The root is contained in a cell-lined sac (the *hair follicle*), which is embedded in the lower layers of the skin. The follicle is a multi-layered unit, which is lined with cells that synthesize the hair protein (called *keratin*) and form the fibrous structure. Other cells add the pigment. How the individual protein molecules are arranged during formation of the hair determines whether hair is straight or curly. Oil glands in the lower layers of the skin and adjacent to the hair follicle secrete oil that serves to maintain the flexibility of the hair. The exterior shaft of a rat hair is illustrated in Fig. 1. The outer surface of an exposed hair is covered by many very thin, nonliving cells which overlap one another. These flat cells may be thought of as the skin of the hair. They serve in part to retain moisture and the oil produced by the oil glands near the hair follicle. Bleaching and other harsh treatments can damage this protective layer. The main body of the hair is a hollow core of very tough protein. This layer contains the pigment melanin, the amount of which is the main determinant of hair color. The structure of this inner layer is altered in a permanent wave. Chemical bonds which maintain the hair structure are broken by chemicals and then allowed to reform while the hair is curled.

Fig. 1 × 185

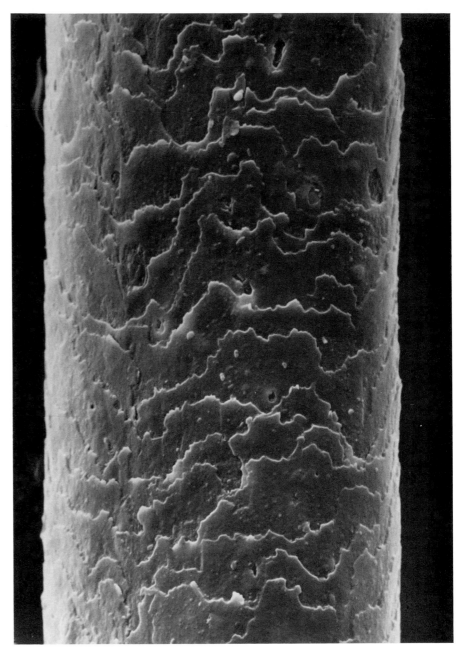

Fig. 2 × 1,385

A high magnification view of a single human hair is shown in Fig. 2. The skin cells mentioned on the previous page are easily seen in this photograph. Vigorous brushing removes dirt from hair but the use of a metal brush often removes these protective surface cells, reducing the ability of the hair to store surface oil. Whereas this might be attractive, this makes the hair more susceptible to mechanical damage and drying. Sometimes a hair will separate from its point of attachment in the hair follicle and fall out of the skin. The follicle then usually becomes active and synthesizes a new hair. Baldness is a result of the inability of a follicle to be activated.

TOOTH

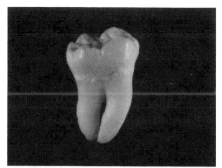

Fig. 1 × 2

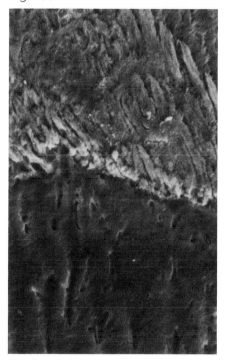

Fig. 2 × 926

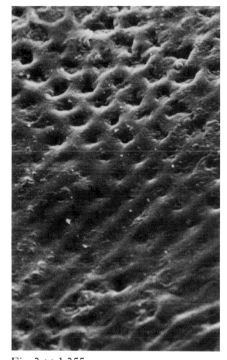

Fig. 3 × 1,355

Fig. 4 × 1,895

Teeth must be able to cut and grind a variety of substances having many degrees of hardness. Thus teeth are designed to be the most rigid component of the body and are covered with one of the hardest substances known, namely, *enamel*. The enamel is located over the exposed portion of the tooth and covers an inner region, which is softer but still quite hard, called *dentin*. Both enamel and dentin are the secretory products of specialized cells called *ameloblasts* (enamel) and *odontoblasts* (dentin). Fig. 1 shows a human wisdom tooth. The upper region seen is the enamel; the roots are covered by dentin. Fig. 2 shows an enlargement of the region at the gum line. The upper material is enamel; the lower material is the dentin. The enamel on the upper surface of the tooth can be seen to consist of many small units (Fig. 3), each of which is the product of a single ameloblast. If the surface of a tooth is etched with acid and then observed at high magnification (Fig. 4), the individual long crystals of the calcium salts can be seen. Most of these crystals are made of the same material as bone. As a tooth matures, the enamel-producing cells die; this accounts for the inability of the body to repair structural damage to teeth.

TONGUE

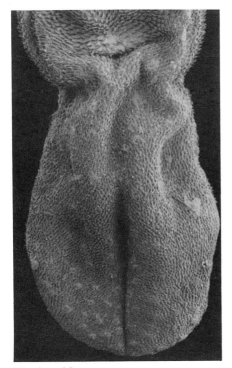

Fig. 1 × 18

Fig. 2 × 252

The tongue is used both for bringing solid food into the mouth, for moving food to the throat and for swallowing, for drinking, and for tasting. The tongue of a rat is shown in Fig. 1. The tongue is a highly flexible muscle that is covered with cells that form an extensively folded surface. The folding produces projections called *papillae*. The papillae vary in shape but most are somewhat pointed, as shown in the enlargement of Fig. 2. An even greater magnification is shown in Fig. 3. The pointed papillae are especially valuable, for example, when licking an ice cream cone. Some of the papillae are shorter and rounder. These papillae have taste buds on their surfaces, which are capable of responding to the few basic tastes, sweet, sour, salty, and bitter. A highly magnified view of a taste bud is shown in Fig. 4. Near the center of the taste bud is a small opening, called a pore. This is shown in the enlargement of Fig. 5. To be tasted, a substance must be in liquid form and must enter the pore.

Fig. 3 × 320

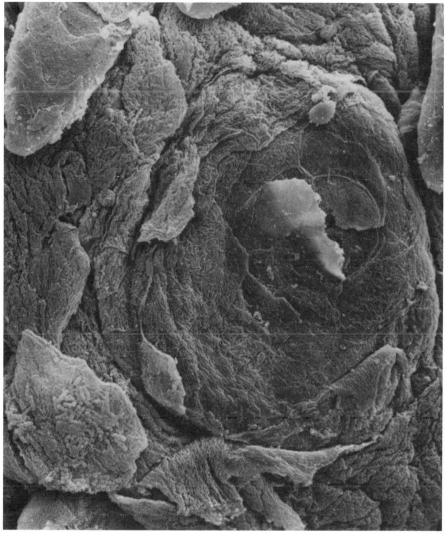

Fig. 4 × 1,582

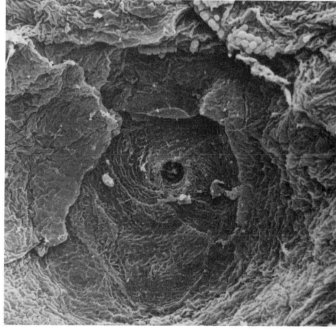

Fig. 5 × 2,610

DIGESTIVE SYSTEM
Small Intestine

The small intestine is a long, muscular tube connecting the stomach and the large intestine. Its principal function is to digest proteins, carbohydrates, and fats, and then to transfer the nutrients to the blood stream. The cells that line the inner surface of the small intestine produce some of the enzymes needed for digestion while others are manufactured in the pancreas and liver and transported to the small intestine through special ducts. To facilitate absorption, the intestine is very long (about 35 feet in an adult human) and the inner wall is highly folded, which increases its surface area. Fig. 1 shows a low-magnification view of a cross-section of the small intestine. The deep folds are easily seen. The units formed by this folding are called *villi*. In a few cases a cut has been made through a villus, which enables one to see its narrow core. In this region masses of tiny capillaries pick up the nutrients.

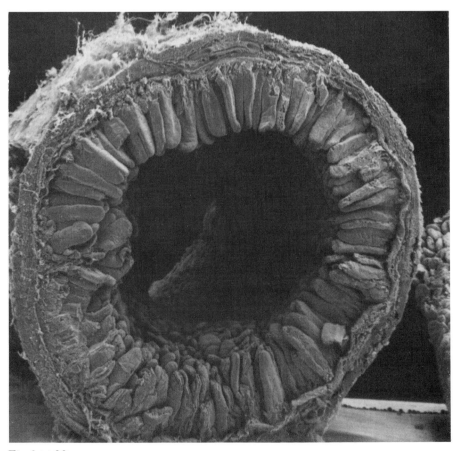

Fig. 1 × 30

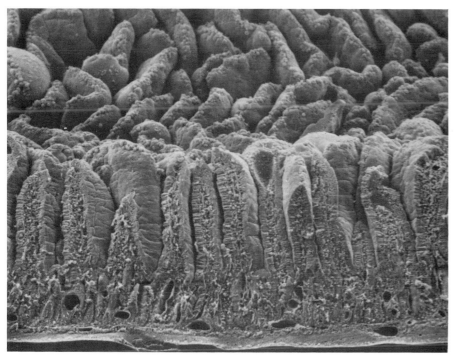

Fig. 2 × 78

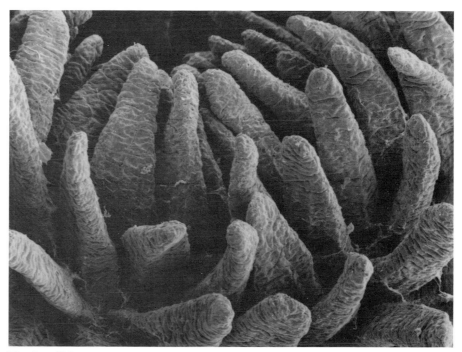

Fig. 3 × 136

An enlarged view of the layer of villi seen in Fig. 1 is shown in Fig. 2. Here it can be seen clearly that each villus is a fold of the inner layer of the intestine rather than a distinct structure. The apparently tangled strands at the base of the layer of villi are part of the fine capillary network into which nutrients flow from the capillaries in the core of each villus. A view looking down on many villi is shown in Fig. 3. Here one can see that the villi are conical and that each villus surface is also folded, which further increases the surface area of the intestinal lining.

Fig. 4 × 244

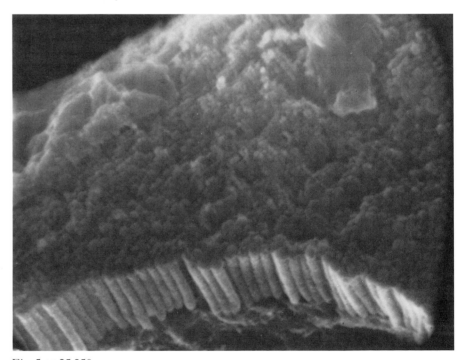

Fig. 5 × 25,250

A higher magnification view of several villi is shown in Fig. 4. It is covered by several kinds of cells, the most numerous of which are *goblet cells* and *absorptive cells.* Goblet cells secrete a mucus that serves to lubricate the inner lining of the intestine so that food will pass through easily and to minimize irritation caused by acid and ingested irritants. In order to increase the surface area of the small intestine further, the surface of the absorptive cells is also highly folded producing finger-like extensions called *microvilli.* Some of these can be seen in the higher magnification photo in Fig. 5.

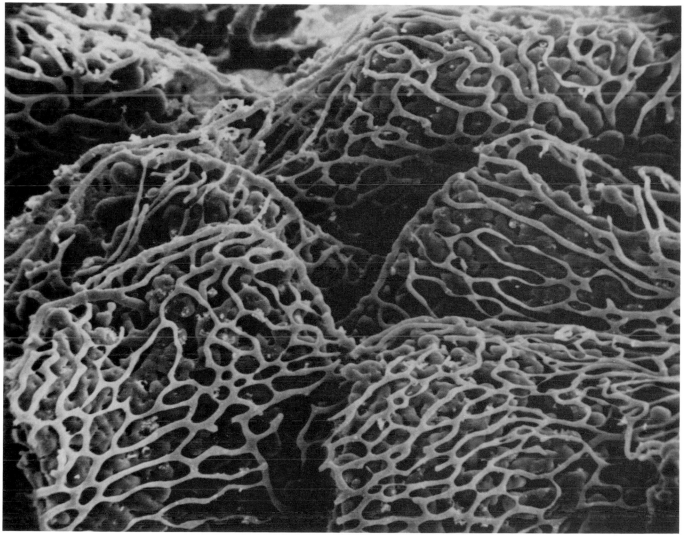

Fig. 6 × 321

An extensive network of capillaries is present in a core region just under the cells covering each villus. The networks of all of the villi join together to form a broad capillary bed underlying the entire villi layer. This network can be visualized in an interesting way. The villi are sliced off the underlayer and all of the cellular material on the surface of the slice is chemically dissolved away. This leaves the capillary bed uncovered. The capillaries are chemically stiffened so they do not collapse and the new surface is examined. A portion of such a network is shown in Fig. 6. Notice how the capillaries criss-cross and branch to provide an enormous surface area so that nutrients can be taken up from the surrounding fluids very rapidly. Deeper in the intestinal wall, the capillaries join to form larger vessels, which in turn form still larger veins and arteries. Fresh blood continually flows into the villi, takes up nutrients, and then brings the nutrients to all parts of the body.

DIGESTIVE SYSTEM
Stomach

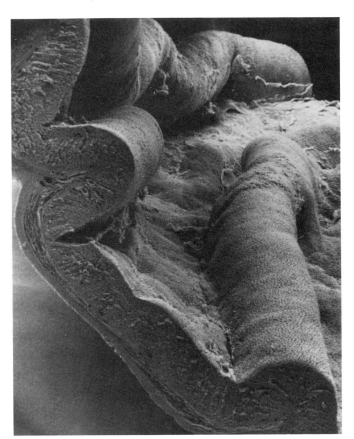

Fig. 1 × 28

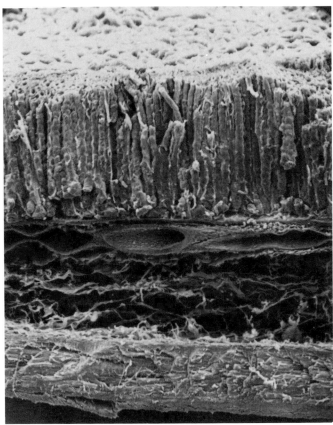

Fig. 2 × 80

The stomach is an enlarged section of the digestive tract. It functions to store food, to digest proteins partially and thereby prepare food for complete digestion, and to absorb salts and excess water. The stomach wall is usually folded when empty (Fig. 1). An enlarged cross-section of the stomach wall shows three layers (Fig. 2). The inner layer consists of many tubular glands which synthesize several substances: a mucus material, an enzyme called *pepsin*, which initiates digestion of proteins, and *hydrochloric acid*, which is needed for pepsin to work. A substance essential for absorption of vitamin B-12 is also produced but by surface cells. Each gastric gland is richly supplied with blood vessels as shown in Fig. 3. When food enters the stomach, powerful muscular contractions churn the contents, breaking up the food into particles smaller than those produced by chewing.

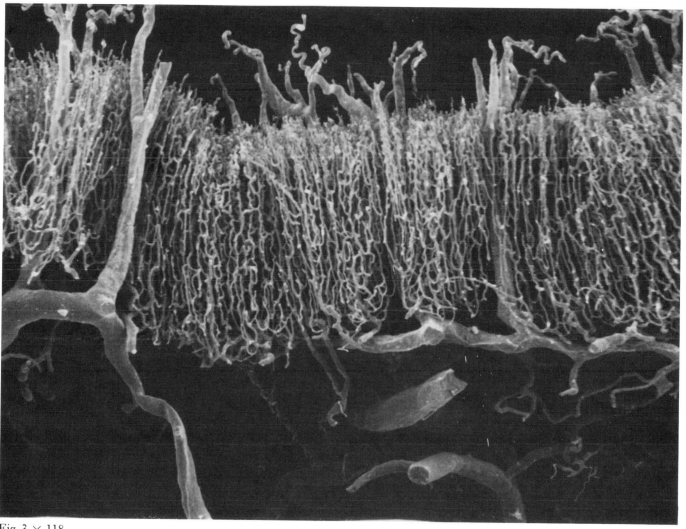

Fig. 3 × 118

RESPIRATORY SYSTEM

Lung

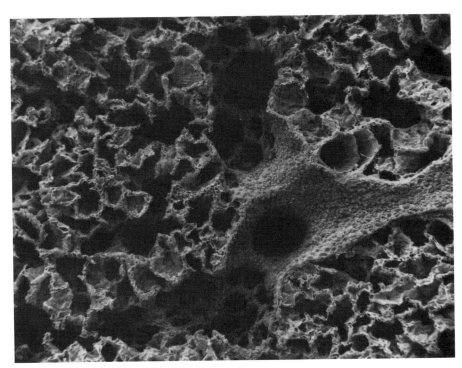

Fig. 1 × 17

Fig. 2 × 156

All cells in animals require oxygen. The principal function of the lung is to bring oxygen from the atmosphere to the blood and to remove carbon dioxide. It does this by providing a mechanical system in which circulating blood is separated from air by a membrane so thin that gas molecules can freely pass through in either direction. Thus the major part of the lung consists of billions of very thin-walled air spaces called *alveoli* around which there are tiny capillaries. The lung is a very porous structure, as shown in Fig. 1, which shows a small portion of a slice through a lung, and Fig. 2, in which a section has been magnified. The number of individual air compartments should be noted. In Fig. 3 the lung tissue is magnified even more in order that one can see the fine structure of the walls of the alveoli. Note the many small circles in the alveolar walls. These are the capillaries through which blood flows. The distance between the blood vessels and the air spaces is little more than the thickness of a single cell so that the air molecules do not have far to go. There are several types of cells in the alveolar wall; one of these, which is covered with pebble-like structures, is shown in Fig. 4. This type of cell is not only a barrier between air and blood but also secretes a detergent-like substance that keeps all fluid in the lungs in the form of a single thin layer; that is, droplets, which would be too thick for efficient passage of gas molecules, do not form. Fig. 5 shows another important type of cell found in the lung, namely, the *phagocyte*. These cells, like similar cells found in the blood, are able to ingest small particles and are responsible for "eating" harmful particulate matter that might be inhaled in the lungs.

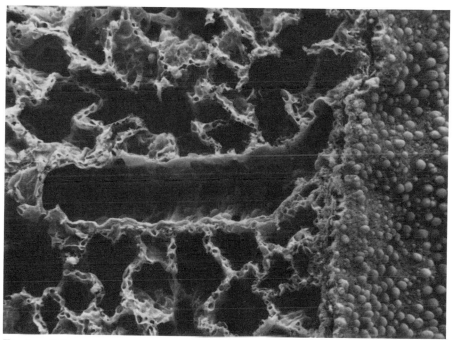

Fig. 3 × 328

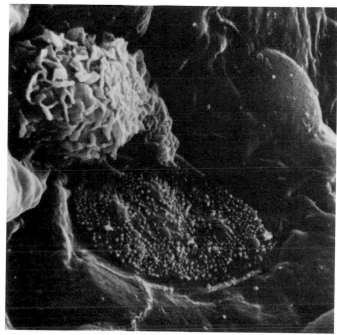

Fig. 4 × 5,500

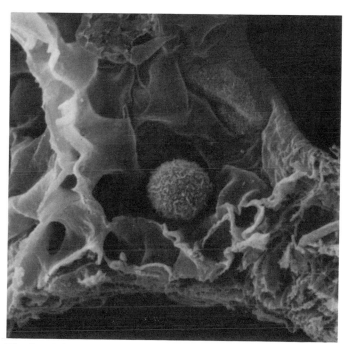

Fig. 5 × 1,630

RESPIRATORY SYSTEM

Cilia

Fig. 1 × 750

Fig. 2 × 2,836

On the previous page a cell whose function is to ingest particulate matter that gets into the lung was described. Here we illustrate another mechanism for eliminating particles from the lung and other parts of the respiratory system. On pages 70–74 we saw that certain protozoans possess tiny hairs called cilia whose beating is a source of locomotion for these free-living unicells. Many of the epithelial cells of the lung, bronchi, trachea (windpipe), and sinuses also contain cilia and these cilia are capable of the same beating motion. Several ciliated cells from the trachea are shown in Fig. 1. The cilia that extend from an individual cell are shown at a higher magnification in Fig. 2. It might be noted that not all of the cells are ciliated. The large cells that surround the ciliated cells secrete a thick, sugar-protein solution called *mucus*. When particles enter the respiratory system, they frequently become trapped in the slimy mucus secretion. Then, nearby cilia are stimulated to beat. Their beating motion is primarily in a single direction, namely, toward the mouth, so that in time the trapped particles reach the throat or mouth, from which they can be eliminated.

LIVER

Fig. 1 × 58

The liver is the largest and most versatile organ in the body. It has many functions among which are the following: synthesis of blood proteins, production of digestive fluids, storage of carbohydrates as an energy reserve, storage of vitamins and iron, removal of many waste products from the blood, and detoxification of harmful chemical compounds. The liver receives nutrient-laden blood directly from the intestine and removes many of the substances just mentioned. In order that all of the blood is able to pass frequently through such a large organ, the liver is subdivided into cell clusters called *hepatic plates*, each of which has a blood supply. A cross-section of the liver is shown in Fig. 1. The small circular areas are blood vessels seen in cross-section. A higher magnification view is seen in Fig. 2. Here one can see that the liver cells are clustered and that there are many channels between these clusters. These channels are actually capillaries that are interconnected. Fig. 3 shows an interesting way to look at the arrangement of these capillaries. The capillaries are injected with a liquid plastic that hardens and then all of the cellular material is dissolved away, leaving the casts of the capillaries shown in the figure. Note that they consist of numerous branches emerging from a single blood vessel.

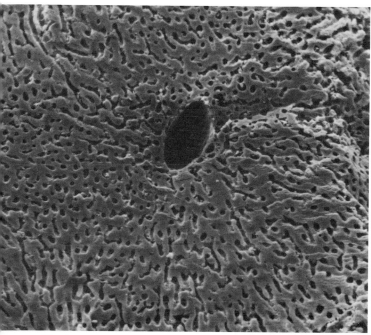

Fig. 2 × 170

Fig. 3 × 98

KIDNEYS

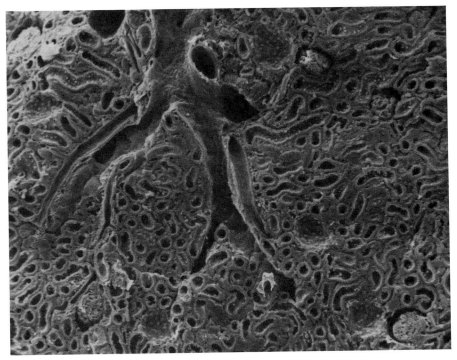

Fig. 1 × 68

The kidneys are a pair of organs situated slightly above the waistline and toward the back of the body. They have two main functions: to remove metabolic waste and unwanted chemicals from the blood and to maintain the correct concentration of salts throughout the body. To accomplish this, the kidney is divided into a large number of multi-tubular units called *renal corpuscles* or *nephrons*. In these units a network of capillaries is in close contact with a network of urine-collecting tubules. In a fairly simple way, water from the blood passes through the walls of the capillaries into special collecting cells carrying with it unwanted soluble substances. Later, water and some salts are removed from this liquid and returned to the blood, leaving behind a concentrated solution of salts and waste products known as urine. The amount of material transported to the urine is exactly that which is necessary to maintain the proper concentration of all substances in the blood. The urine is at this point in tiny tubules which join other tubules to form large urine-collecting tubes which drain into a major vessel that carries the urine to the bladder. The kidney can be subdivided into an inner and an outer region. The inner region consists mostly of the larger collecting tubes. The outer region, which is called the *cortex*, is shown in cross-section in Fig. 1. It consists almost entirely of blood vessels and the primary urine-collecting tubes, known as *uriniferous tubules*. These tubules are the small cavities seen in cross-section in the figure. The large open regions are blood vessels. An enlargement of the tubules is shown in Fig. 2. In the lower right portion of the figure is a small globular cluster. This is the workhorse of the kidney, the *renal corpuscle*. It is in this unit, of which there are several million in each kidney, that the actual transfer of water and salts occurs. An enlarged view of a cross-section of a renal corpuscle is shown in Fig. 3.

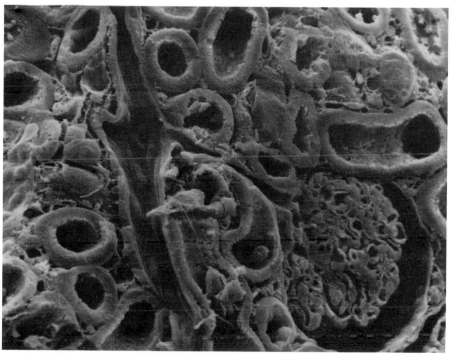

Fig. 2 × 475

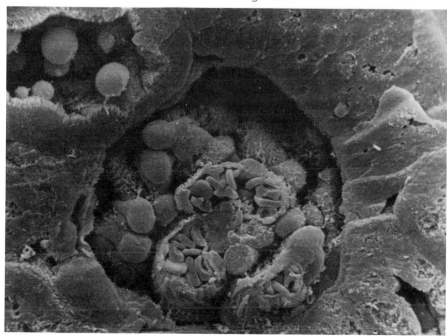

Fig. 3 × 1,176

135

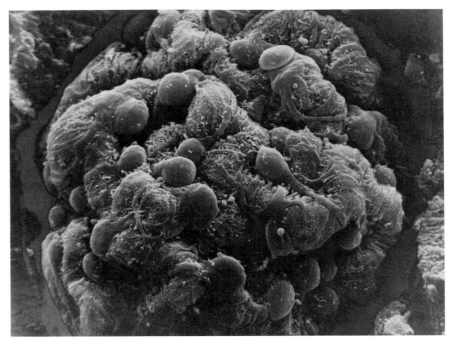

Fig. 4 × 1,840

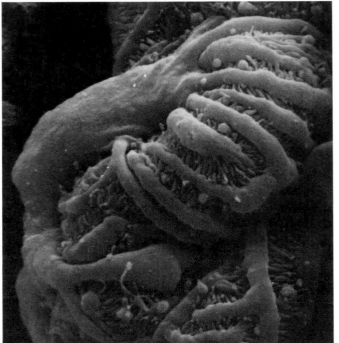

Fig. 5 × 5,157

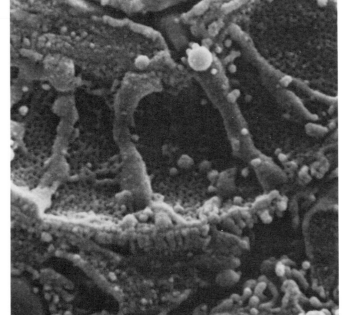

Fig. 6 × 13,700

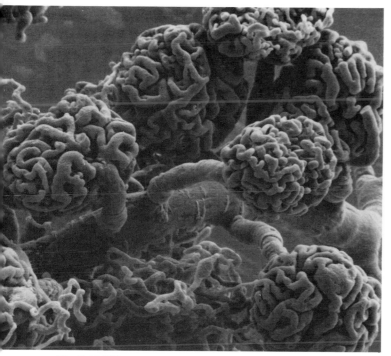

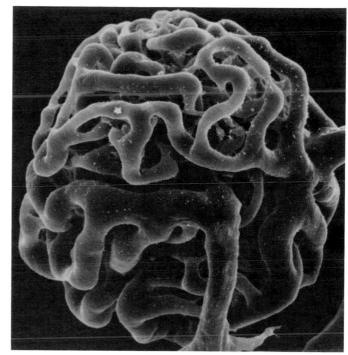

Fig. 7 × 300

Fig. 8 × 605

How a renal corpuscle is organized can be best seen by examining its surface. Such a view is shown in Fig. 4. What is seen in this figure is a twisted capillary. Note that strewn over the surface of the capillary are many multiply branched bodies. These are highly specialized cells whose unusual shape makes them look unlike any other type of cell that we have seen. These cells, which are called *podocytes,* look like an octopus with hundreds of tentacles clinging to the capillary. A single podocyte bound to the capillary surface is shown at higher magnification in Fig. 5. The podocyte with its many arms is in intimate contact with the capillary. Notice how the arms of the podocyte are broken down into very small branches at the points of contact. This increases the surface area of the capillary with which the podocyte is in contact. Liquid from within the capillaries accumulates in the fine slits between the very small branches and is transferred to the uriniferous tubules. Liquid is able to leave the capillaries because the capillaries contain very fine pores. This is shown in the very highly magnified view of the inside of a capillary (Fig. 6). The numerous tiny pores can be seen easily. The highly coiled capillary is called a *glomerulus.* The extent of the coiling can be seen in Fig. 7. This photo was obtained by filling the capillaries with a fluid plastic material, which was then allowed to harden. After hardening, the entire unit was treated with a strong alkali, which removed all cellular material including the walls of the capillary. What remains is a cast of the capillary, namely, the hardened plastic. An enlargement of one of these plastic casts is shown in Fig. 8.

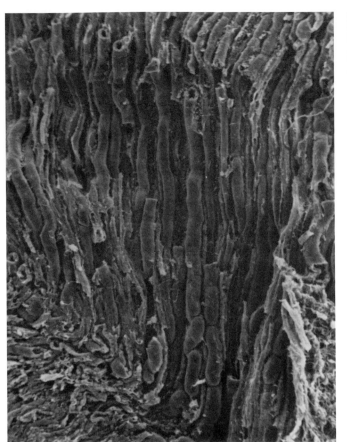

Fig. 9 × 103

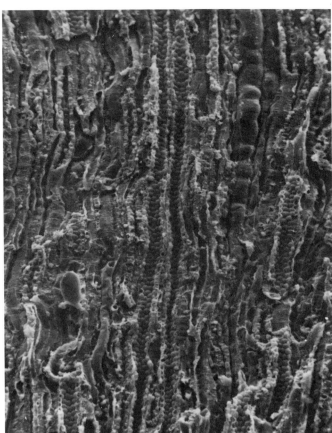

Fig. 10 × 180

Once the dilute urine leaves the glomeruli and passes by the podocytes, it is collected in the tubules described earlier. Figs. 9 and 10 show a region of the kidney in the inner reaches of the organ. This region consists primarily of collecting tubules. Several different types of tubules, as well as small blood vessels, are seen in these figures. Many changes occur in the urine as it passes through these tubules. For example, both water and some salts, which are not to be excreted, are resorbed by the many small blood vessels that surround the tubules.

NERVES

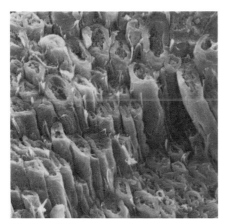

Fig. 1 × 900

Fig. 2 × 1,320

The *neuron* or nerve cell is the structural unit of the nervous system. Neurons generate electrical impulses in response to different stimuli such as heat, pressure, and cellular chemicals. A neuron can also receive electrical impulses from another nerve cell. The receivers are short branched hairlike appendages called *dendrites*. The dendrites transfer the impulse to the main body of the cell which in turn acts as a transfer point, shuttling the impulse to a very long thin branch of the neuron called an *axon*. The axon is the "electrical wire" along which the impulse moves. In nerve fibers the end of an axon terminates on the dendrites of the next neuron along the path of the fiber. It may also contact a muscle cell in which case the electrical activity of the nerve impulse is converted to mechanical activity. The main nerves of the body consist of axons and dendrites of many cells closely packed together like cables in a wire. This arrangement can be seen in the two figures. The small figure shows a cluster of nerves viewed at an angle to the axis of the fiber. Each of the elongate structures is an axon. The large figure shows a cross-sectional view of a fiber. Most of the axons shown in this figure are wrapped in a fatty substance called *myelin*, which serves as an electrical insulator.

EYE

The eye is a spherical organ which receives light, converts the image to electrical signals, and transmits these signals to the brain. The light-sensitive region is a multilayered structure called the *retina*, which is located on the back inner surface of the eye. A cross-section of the retina is shown in Fig. 1. The elongate cells at the upper part of the figure are the light-sensitive cells, which are called *rods* and *cones* because of their shapes. These cells contain pigments which absorb light and initiate a chemical reaction, which in turn generates a weak electrical signal at the lower end of the cell. This signal is received by the larger round cells seen in the figure just under the visual cells. These cells amplify the signal and transmit it to nerve cells. The long appendages (*axons*) of each nerve cell are grouped together in a bundle called the *optic nerve* and carry the electrical signal directly to the brain where the signals are converted to the image we know of as vision. The light-sensitive pigment is converted to a light-insensitive form after an image has been received. The lower portion of the figure shows a layer that consists of cells whose function is to restore the visual pigment to the sensitive form. This layer also contains blood vessels which provide nourishment for all of the cells in the retina.

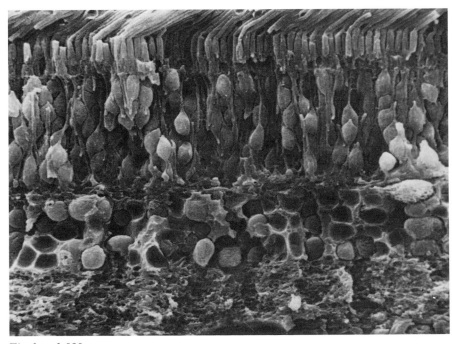

Fig. 1 × 1,090

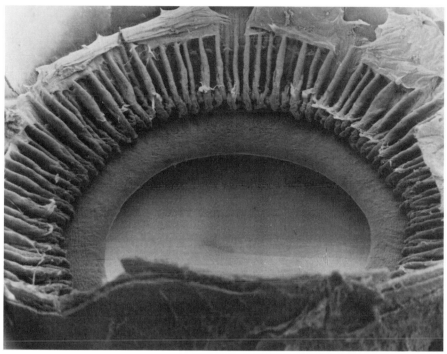

Fig. 2 × 19

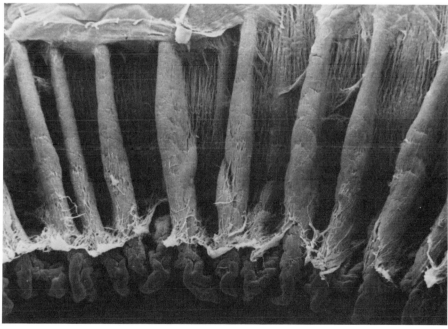

Fig. 3 × 74

To protect the eye against extremely intense light yet to gather as much light as possible in dim light, the eye regulates the size of the pupil, the dark opening through which light enters the eye. This is accomplished by the *iris,* a colored ring-shaped muscle that surrounds the pupil. A cross-section of the iris is shown in Fig. 2. The central clear area is the pupil. The iris consists of a set of short muscle fibers (also shown enlarged in Fig. 2) attached to a stationary outer region and a flexible smooth inner ring. Contraction of these muscular fibers pulls the ring outward, enlarging the pupil. Bright light prevents contraction, keeping the pupil small.

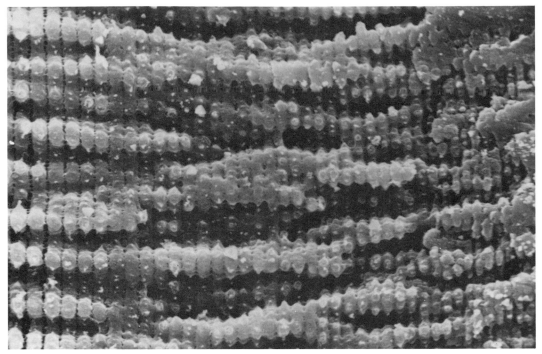

Fig. 4 × 1,520

Fig. 5 × 3,180

Although the major focusing effect is caused by the curved front surface of the eye, the ability to vary the focal point is controlled by a lens, which is located just behind the pupil. The lens is a transparent object whose shape can be altered slightly to change its focal point. To the naked eye, the lens appears to be a blob of jelly. However, the lens actually consists of numerous closely packed elongate cells called *lens fibers*. Two types of these cells are shown in Figs. 4 and 5. Note in Fig. 5 how adjacent cells are linked together by a multitude of ball-and-socket-like joints.

EAR

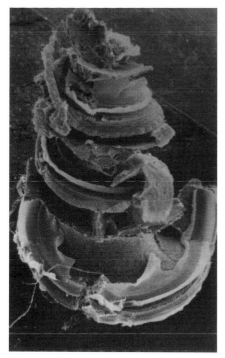

Fig. 1 × 30

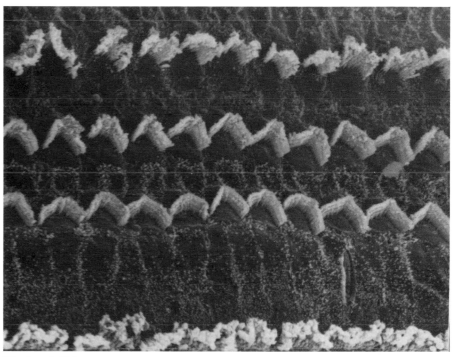

Fig. 2 × 1,916

The ear is the organ of hearing. It consists of three regions: the external ear, which is the visible part, the middle ear, and the inner ear. The external part is designed to catch the vibrations of air that we call sound. The sound enters the passage into the head and sets up vibration of a membrane called the *ear drum*. This membrane vibrates along with the air. The ear drum separates the ear canal from the middle ear. Attached to the ear drum are two important systems. One is a set of liquid-filled tubes, which are used for maintaining balance; the level of the liquid in these tubes informs the brain whether the body is vertical, horizontal, etc. If this liquid sloshes around due to violent motion, the signals get confused and one feels dizzy. The other system is a set of very fine bones called the *hammer, anvil,* and *stirrup,* which transfer vibrations of the ear drum to the inner ear, in which the main sound sensor is located. The inner ear contains a complex and highly inner-vated organ called the *cochlea.* This organ, which has a spiral shape, is shown in Fig. 1. Rows of cilia within the cochlea (shown in Fig. 2) pick up the vibrations from the bones of the middle ear, and convert them to signals used by the nerves to transfer the message of sound to the brain.

OLFACTORY RECEPTORS

The olfactory receptors are responsible for the sense of smell. The ability to detect smells is extraordinarily sensitive. No manmade detector of chemicals can pick up, let alone identify, as small a number of molecules as can the nose of most animals. The mechanism of smell is exceedingly complex and is one of the most poorly understood biological phenomena. In principle, the phenomenon is much like sight (see page 140) in that receptor cells respond to chemicals and produce an electrical impulse that is amplified and transmitted to the brain. In contrast with the cells responsible for vision, the olfactory cells are in direct contact with the environment; in fact, they are the only nerve cells in the body that do make such direct contact. The olfactory cells are organized as tissue in the nasal passages over which air passes. Fig. 1 shows some of this tissue, which appears as a mass of fibers. A closer look, shown in Fig. 2, shows that the fibers are actually hairlike appendages emerging from a bulb at one end of an elongate cell; the main body of the cell is not seen in the figure. These fibers are kept moist by glands beneath the surface. Gas molecules dissolve in the liquid and bind to specific receptors on these fibers. This generates an electrical impulse which is carried by the olfactory nerve to the brain.

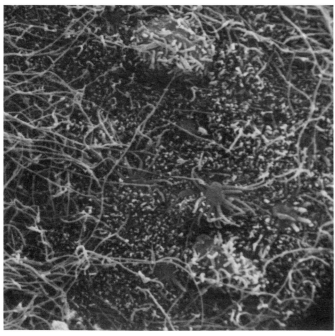

Fig. 1 × 3,820

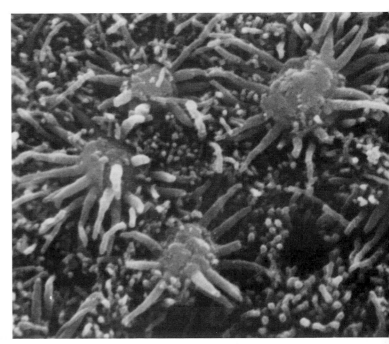

Fig. 2 × 8,970

144

TESTIS

Fig. 1 × 39

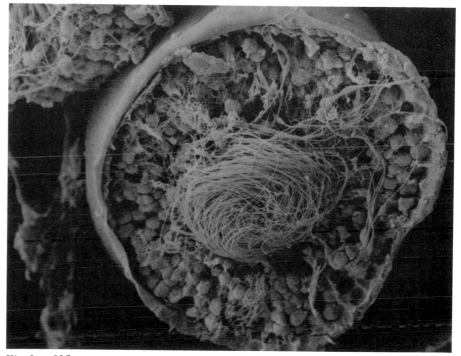

Fig. 2 × 335

The *testis* is the part of the male reproductive system responsible for the production of the male sex cells or *sperm*. The testis consists of many highly folded tubes called *seminiferous tubules*, in which sperm form. A cross-section of several of these tubules of the testis of a rat is shown in Fig. 1. The enlargement (Fig. 2) shows that each tubule consists of a solid wall lined with a variety of cell types. The hollow core of this tubule is filled with many mature sperm cells. The tangle of fibers are the sperm tails. Individual rat sperm cells are shown in Fig. 3. Note that each cell has a small hooked head and a long tail. The head contains all of the genetic material; the tail is used for locomotion when the sperm swims to the egg. Sperm are made in special sections of the testis. One of these units is shown in Fig. 4. Many of the cells adjacent to the walls of the sections are cells destined to become sperm cells. Cell division occurs at the periphery of these units pushing the maturing sperm toward the center. Many sperm tails can be seen in the central portion of this sperm-producing unit. Fig. 5 shows an enlarged area. The newly matured sperm are at the lower part of the figure. Several recently divided precursor cells can be seen elsewhere in this figure.

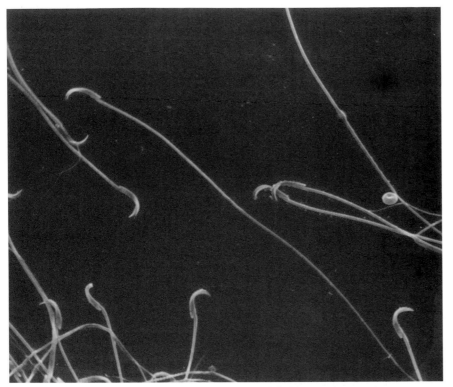

Fig. 3 × 785

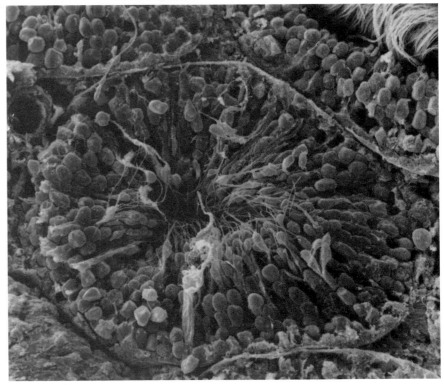

Fig. 4 × 366

146

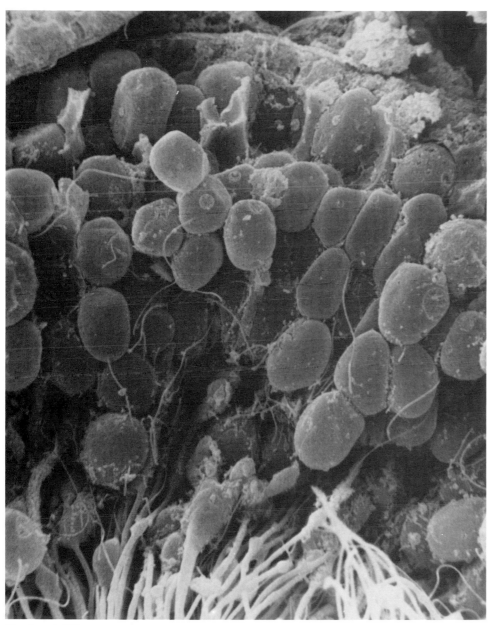

Fig. 5 × 1,370

OVIDUCT

Fig. 1 × 72

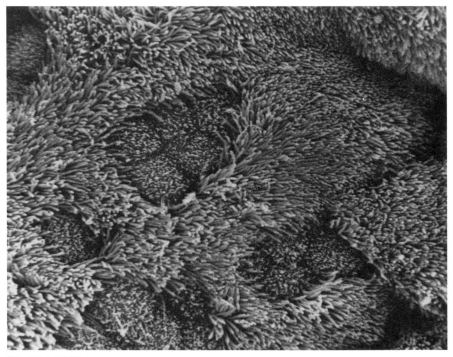

Fig. 2 × 3,225

The oviducts are two tubes responsible for transporting a newly-formed egg from the ovary to the uterus. In humans they are also called the *Fallopian tubes.* Each oviduct has muscular walls whose contractions and relaxation cause wavelike ripples that propel the egg toward the uterus. A cross-section of a portion of an oviduct is shown in Fig. 1. The oviduct wall is highly folded. The inner surface is lined with columnar cells seen easily in the lower part of the figure. Some of these cells are heavily ciliated; an enlarged view of the surface of a ciliated cell is shown in Fig. 2. The cilia beat both toward and away from the uterus and their function is not yet known. The oviduct also contains many nonciliated cells. These cells synthesize and secrete substances which are believed to be nutritious for the egg. This is valuable for the egg often takes three or four days to pass through the tube.

UTERUS

The young animal develops in the uterus of an adult female and is expelled from it at the time of birth. The uterus has a thick muscular wall, whose muscles are among the strongest in the body. A cross-section of a uterus is shown in Fig. 1. The inner layer, called the *endometrium*, is a soft blood-rich layer containing many nooks in which an egg can become lodged. The many tiny openings on the endometrium are the ducts of uterine glands. One such opening is shown in the higher-magnification view of Fig. 2. When an egg is fertilized, a process begins called *implantation*, in which the egg becomes lodged in the endometrium. The uterine glands respond to fertilization and implantation by secreting a nutrient-rich fluid which nourishes the developing egg until a placenta forms. If fertilization does not occur by a certain time in the menstrual cycle of a female, much of the endometrium detaches from the muscular layer and is progressively shed and expelled over the period of a few days. A new endometrium then develops for a second attempt; this is ready by the time a new egg enters the uterus. The uterus grows during pregnancy. At the time of birth, contractions of the heavy muscular wall shown in Fig. 1, which has grown considerably in strength during pregnancy, force the newborn out of the uterus into the birth canal.

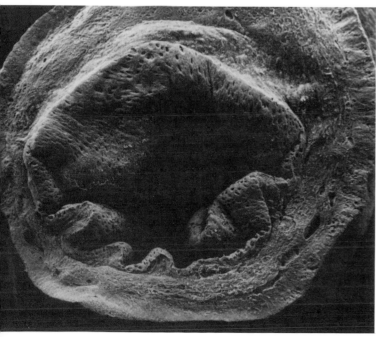

Fig. 1 × 36

Fig. 2 × 2,300

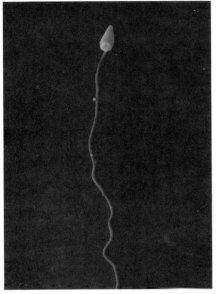

Fig. 1 × 765

Fig. 2 × 2,335

SPERM AND EGG

Fertilization

Most organisms produce *germ cells*. The cell produced by the male is the *sperm*; that produced by the female is the *egg*. Each contains one copy of all of the genetic information of the parent organisms. However, sperm and eggs are very different types of cells—both visually and functionally. For example, the egg of all organisms—in this case, a sea urchin (Fig. 1)—is a large sphere, whereas sperm, shown both on the egg in Fig. 1 and at higher magnification in Fig. 2, are much smaller than the egg and have two definite parts. The genetic material of the sperm is contained in its tiny arrow-like head. The long tail is structurally similar to the cilia in microorganisms seen earlier in this book and is the means of propulsion of the sperm. In almost all organisms the egg is delivered to some resting place and there it stays while sperm seek it out. The sperm whip their tails violently and propel themselves toward the egg. In animals in which fertilization occurs within the mother, the motion is definitely nonrandom as all of the sperm deposited in the female move primarily in the direction of the egg, responding to signals which are not yet known. However, the route in such animals is long so that a billion or more sperm are deposited in the female and only a few hundred may actually reach the egg. In marine animals in which unfertilized eggs are laid and sperm are deposited nearby in the water, the motion of the sperm is more random and collision between sperm and egg probably occurs by chance. Fig. 3 shows a magnified view of a sperm that has found an egg. Typically its tail lies along the surface of the egg and its head is poised for entry. Usually, up to 100 sperm will come into contact with an egg. By time-lapse photography it is possible to see each of them flapping its tail violently, trying to drive its head through the wall of the egg. Finally, one of them succeeds (Fig. 4); notice how the wall of the egg is being dissolved away as the head enters. Interestingly, as soon as one sperm gains entry, the frantic motion of all others ceases.

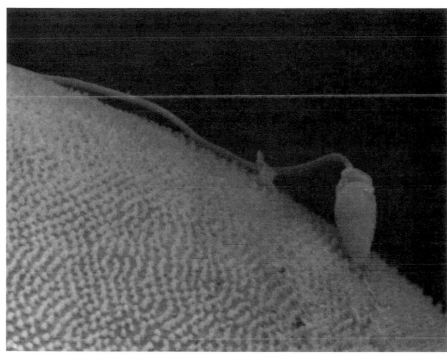

Fig. 3 × 10,285

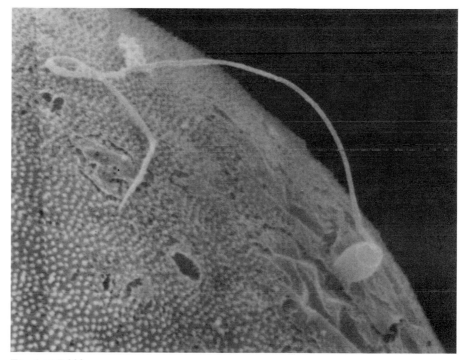

Fig. 4 × 7,800

DEVELOPMENT
Zebrafish

In the previous pages we saw how an egg is fertilized. Now we follow the subsequent development of a fertilized egg of the zebrafish, a common freshwater fish found in home aquariums. The zebrafish lays eggs, which are fertilized by sperm that have been released to the surrounding water by a male. Several minutes after fertilization, the contents of the cell start to move and churn around; ultimately, there is some order to this movement and all of the non-yolk material in the egg moves to a single region on one side of the egg. This region, which is called a *blastodisc*, appears as a swelling on the surface of the egg. This swelling can be seen on nearly the entire front surface of the egg shown in Fig. 1. Soon afterward, a depression forms along the midline of this bulge (Fig. 2). This depression is called the *first cleavage line* and indicates that two cells will soon form on both sides of the depression. The bulge of the blastodisc increases in size slightly and the cleavage line becomes deeper and soon two cells are easily seen. Each of these cells enlarges and a second cleavage line appears, which is perpendicular to the first cleavage line. This can be seen in Fig. 3, in which four cells are evident. Each of the four cells continues to grow and finally two more cleavage lines appear that divide each cell into two parts. At this point, shown in Fig. 4, there are eight cells, arranged in two rows of four cells each. During all of this time, growth of the cells

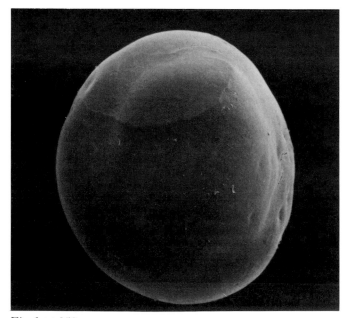

Fig. 1 × 150

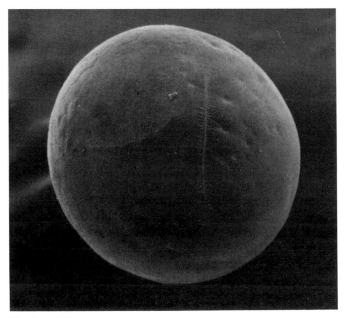

Fig. 2 × 143

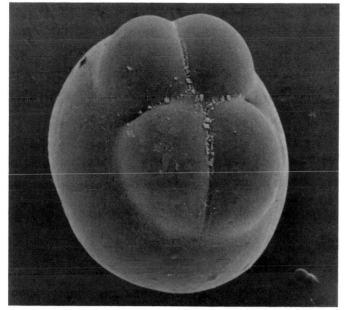

Fig. 3 × 135

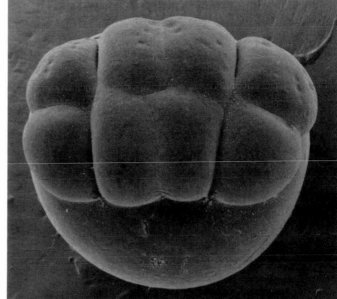

Fig. 4 × 148

Fig. 5 × 120

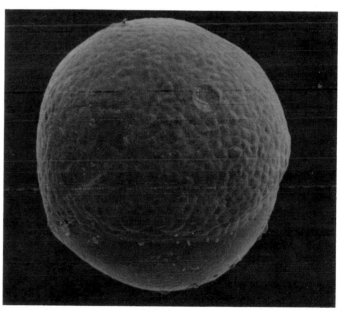

Fig. 6 × 140

has utilized the nutrient of the yolk sac. Thus, the fraction of the total volume that is the yolk sac decreases (compare Figs. 2 and 4). Division continues but without very much increase in total volume of the cells. Thus, when the stage shown in Fig. 5 is reached, the egg is capped with a large cluster of fairly small cells. Notice the small swelling at the junction of the cellular mass and the yolk sac, where new cells are being formed over the yolk region. In time, the yolk is enclosed in a nearly spherical mass of small cells of fairly uniform size. This is shown in Fig. 6. Notice how the cells no longer project up from the surface of the sphere. At this point, an important change occurs in the structure of the egg. That is, a small spherical depression appears on the surface of the egg. This is easily seen in Fig. 6. Development of the embryo is determined by cells in this region.

A thickening now appears on the surface of the egg; this is called the *embryonic band,* for the embryo will develop directly from this ridge. The ridge increases in size as more cells migrate into this region. This generates the structure shown in Fig. 8, which is beginning to resemble an animal with a head and tail wrapped around a yolk sac. The animal is considered to be an embryo at this stage. With continued development, as shown in Fig. 9, the fish-like form becomes more evident. The tail has grown longer and is terminated by a flat section that will be the tailfin. Bumps on the head will develop into eyes and pectoral fins. The embryo is still totally dependent on the yolk for nourishment. (A small amount of yolk has been removed and is shown in Fig. 10. Note how the yolk is organized into crystalline grains.) About four days after fertilization, the egg hatches and the animal is released. Interestingly, the newborn animal is incompletely developed, as shown in Fig. 11.

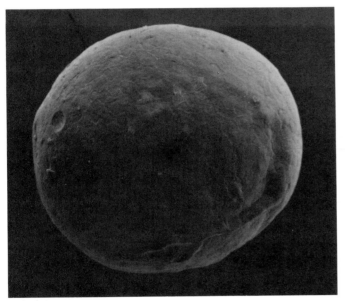

Fig. 7 × 140

Fig. 8 × 85

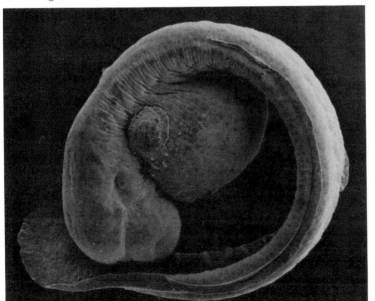

Fig. 9 × 105

Fig. 10 × 1,000

154

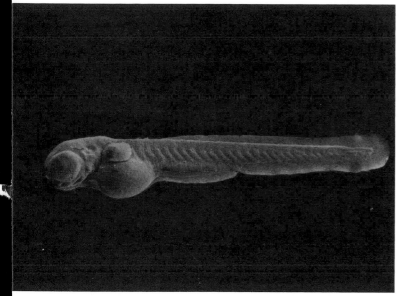

Fig. 11 × 37

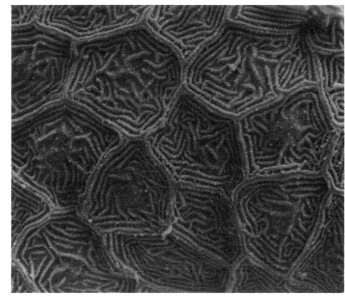

Fig. 12 × 2,150

The eyes are still large bulges incapable of vision, the yolk sac remains, no fins have yet formed, and the fish lacks color. The young fish swim quite poorly and in nature most of them are quickly eaten by other fish. Probably no more than one to ten fish in a million survive more than an hour after birth. During the next few days the fish continues to develop. In so doing, the yolk sac is completely absorbed but by this time the fish is mature enough that it is able to gather food from its environment. Zebrafish are lovely streamlined fish capable of swimming very rapidly. Their name comes from prominent narrow black stripes running along the body. A close look at the skin of a zebrafish shows beautiful microscopic patterns on its skin cells, as shown in Fig. 12.